# Space and Oceans

# Space and Oceans

## Tracking Plastic Pollution in the Arctic Ocean from Space

**Final Report**

Adelaide
2022

*International Journal of Space Studies*
Volume 3 Number 2, 2022

The MSS 2021 Program of the International Space University (ISU) was held at the ISU Central Campus in Illkirch-Graffenstaden, France.

The cover image represents Jökulsárlón, a Glacier Lagoon located in the southern part of Vatnajökull National Park, Iceland. The man on the inflatable boat symbolises the path towards a sustainable future for the Arctic Ocean.

Cover page image source: Copyright Credit ©unsplash.com/rolfgelpke
Edited by:      Gabriel Bueno Siqueira
Layout:         Extel Solutions, India
Font:           Minion Pro

ISBN's
Soft cover:978-1-922737-49-6
Hard cover: 978-1-922737-50-2
E-book: 978-1-922737-51-9
PDF: 978-1-922737-52-6

Published and edited by

Making a lasting impact
An imprint of the ATF Press Publishing Group
owned by ATF (Australia) Ltd.
PO Box 234
Brompton, SA 5007
Australia
ABN 90 116 359 963
www.atfpress.com

# Acknowledgments

Throughout the writing of this report, the team has received a great deal of support and assistance.

The team Space and Oceans would first like to express their deepest appreciation to Professor Doctor Bertrand Goldman and Professor Doctor Hugh Hill, our TPR Faculty Interfaces, for their patience, enthusiastic guidance and useful critiques of this report. The team would like to extend their deepest gratitude to Prof. Walter Peeters, whose expertise was invaluable in formulating the focus and approach of the project. Your insightful feedback pushed the team to sharpen their critical thinking and brought this report to a higher level. The team would also like to thank for their precious help all the members of the faculty and staff of the International Space University, in particular:

Danijela Stupar, Research Associate in Remote Sensing
Doctor Taiwo Tejumola Raphael, Space Systems Engineer
Prof. Chris Welch, Masters Program Director
Prof. Gongling Sun, Space System Engineer and Policy Analyst
Muriel Riester, Library Manager
Tommaso Tonina, Teaching Associate
Juan de Dalmau, ISU President.

Furthermore, the team is extremely grateful to Silje Bareksten and Doctor Su-Yin Tan for proposing the subject of the team project and for sharing their expertise. In addition, the team would like to send special thanks to the following external experts who have contributed to the success of this work with their time and wise counsel.

**External experts:**

Marcello Ingrassia, Psychologist, Human Scientist, and Team Building Consultant, Inwind, Italy Vera Pinto Gomes, Policy Coordinator, European Commission, Belgium Sylvain Michel, Innovation and Operations and System Engineer, CNES, France Albert Font de Rubinat, Co-founder and sustainability manager, Good Karma Projects, Spain Jordi Oliva, Co- founder and business developer, Good Karma Projects, Spain Mark Cashen, Sustainable restaurateur, As One, Ireland Emir Sirage, Chief Operations Officer, AIR Centre, Portugal Pedro Silva, Chief Technology Officer, AIR Centre, Portugal Jose Moutinho, Chief Network Officer, AIR Centre, Portugal Sonja Behmel, CEO and co-founder, WaterShed Monitoring, Canada Olanike Maria Buraimoh, Lecturer in Environmental Microbiology, University of Lagos, Nigeria Federico Rondoni, ISU alumnus, MSS20, Italy Ludvic Pagé-Laroche, Project Manager, WaterShed Monitoring, Canada.

# Abstract

There is a growing concern over the ubiquitous distribution of plastic pollution that is evolving in the Beaufort Gyre in the Arctic Ocean, prompting international collaboration and new environmental measures. Despite an exponential increase in the amount of information on plastic in the five ocean gyres using space technologies (i.e. satellites) and applications, data specific to trends in the Arctic Ocean remain scarce, requiring innovative solutions to monitor the growing situation. The geophysical characteristics and presence of sea ice make it difficult for current technologies, specifically Earth observation, to detect and track microplastic. Marine pollution is recognised as an immediate threat to both land and marine ecosystems. Satellites have proven useful in identifying ocean plastic patches and current movements in other oceans but little research has been applied to the Arctic, a region that impacts eight countries making up the Arctic Circle. This is in part because plastics are less observed compared to melting ice caps. Nonetheless, as noted by environmentalists and academics, waiting until an ecological problem becomes a disaster only exacerbates the situation. Moreover, as sea ice melts, new economic opportunities for marine activities including new trade routes (i.e. Transpolar Sea Route), fishing locations and resource exploitation will lead to increased pollution. Due to the expanse of the Arctic Ocean, its frozen characteristics and developing plastic problem, adapting current space technologies and applications to monitor and track plastics on the surface, in the ice and below the surface could be a solution for tackling the problem before it meets the level of the existing five gyres. This interdisciplinary team project paper investigates the use of Sentinel-2, Sentinel-6, Fourier-Transform

Spectroscopy, stratospheric balloons and autonomous underwater vehicles to provide an integrated strategy, including communication and outreach, to tackling marine plastic pollution while recognizing that it is necessary to also prevent plastics from entering the ocean in the first place. The general cost of the operations was considered in tandem with an overview of the financial loss and societal burden of ocean plastic. Included in the paper is a proposal for the enhancement of the Arctic Council using scientific diplomacy and best practices for marine debris management and recommendations for future research.

# Faculty Preface

*'You cannot get through a single day without having an impact on the world around you. What you do makes a difference, and you have to decide what kind of difference you want to make.'* - Jane Goodall, *Primatologist.*

In Strasbourg, France, in October 2020, a spirited team of fifteen young people of the Master of Space Studies (MSS) program sought to have an impact on the world around them. Their MSS team project (TP) odyssey took them from the leafy banks of the Rhine River, to the tropical waters of the Lagos Lagoon, to the frigid depths of the Arctic Ocean. It took them from the sophistication of Earth observation satellites and data reduction, to the revolution of underwater robots, to the uncertainties of new territories and new treaties, to the complexities of climate and feedback loops.

The pollution of our oceans by plastic concentrates in gyres located thousands of kilometers from home and yet we find it, literally, on our plates. Megatons of plastics are released into the briny waters annually via our grocery bags, fishing nets, lost shipping containers, abandoned face masks and single-use bottles. Plastic has entered every aspect of our lives and makes our tools lighter and stronger, but this very durability returns to haunt us. The hardy polymers last much longer than the usage we make of them. With time and the elements, plastics abrade and degrade into contamination for ocean plankton and, ultimately, the fish on our plate.

Undeterred by this challenge, the team tackled the problem comprehensively, delivering a wide- ranging Literature Review in late December, 2020. Thereafter, and following discussion, debate and reflection, they focused their research on aspects of Arctic pollution and associated, Space-related mitigation strategies.

The team navigated the high seas of research, team work, and dead-lines, and remarkably did so without leaving home. For the first time in ISUs history and on account of the global pandemic, our modern Ulysses realised their odyssey remotely: discussing and debating, analyzing data, interviewing experts and solving problems. Their voyage exposed them to the essence of teamwork: self-organization and discipline, listening and communicating with colleagues, researching completely new topics, respecting differences of opinion, and maintaining group morale.

This Final Report marks the culmination of their odyssey and represents the fruit of their labors. It has been a pleasure working with them and observing their scholarly progress and overall development. For sure, we congratulate the team and wish them the very best for the remainder of their time at ISU and beyond!

Associate Professor Bertrand Goldman and Professor Hugh Hill,
Faculty TP Interfaces,
ISU Central Campus,
Strasbourg,
France.

# Authors' Preface

Our team consists of fifteen students from eight different countries taking part in the Master of Space Studies Program at the International Space University (ISU). Our backgrounds encompass various sectors, notably engineering, medicine, chemistry, humanities, computer science, archaeology, geology, economics, environmental policy and international relations, and all equally passionate about space; creating a dynamic interdisciplinary framework. The team worked together to find a common objective in the project and to outline a strategy that could lead to its achievement.

One of the challenges the team had to overcome was remote working due to the COVID-19 pandemic which meant that we could not meet in-person. Nonetheless, this made for a unique opportunity to maximise efficiency in a remote-working context. Virtual meetings, email and messaging applications were tools that helped us maintain a continuous line of communication, a key element to working as a team and to foster group spirit. Having a common strategy and setting internal deadlines helped us stay on track throughout these weeks.

The project proposal posed an obstacle for us at first. We had heard a lot more about plastic pollution in the oceans but approaching the problem from a space sector perspective did not seem evident at the outset of the project. Ergo, we decided to start our research from scratch: what is the origin of plastic pollution, in other words how does it end up in the oceans? From that point we could then identify the main plastic pollution routes and research how we could precisely monitor these transportation routes. During the research for our literature review, we learned about the causes and consequences of pollution in the oceans which supports the fact that additional research on solving the pollution crisis is appropriate , urgent and necessary for global health. The many barriers to monitoring ocean debris that we had spotted confirmed that a space-based approach would be necessary to tackle this problem.

However, the challenge at this point of the project was to create an original proposal without only providing mere description of such technologies. As such, we decided on a scope of the project that would address the situation adequately. Our first idea was to build a case study around the Rhine River since we learned through our literature review that rivers contribute a lot to the plastic pollution in oceans. After further examination and discussions with the ISU faculty, we decided on the Arctic Ocean as our case study. This region left relatively unexplored is increasingly threatened by plastic pollution. Thus, our proposal to investigate how to monitor plastic in the Arctic Ocean also serves as a key link to the space applications and activities, and will eventually be the development of an integrated strategy to monitor the ice-covered surface, through innovative technologies based on space and near-space applications.

There are two major outcomes of this project: first, identification of useful space based technologies and applications for tracking and monitoring ocean plastics; second, the proposal for an integrated space, air, and underwater strategy to monitor plastics year round. We hope that this message will reach as many as possible, so that additional investment is put into the space sector for the long term benefit for humankind. We aspire for this work to further the discussion on using space technologies with an interdisciplinary framework and applied to solving the plastic pollution problem.

# Contents

# List of Figures

# List of Tables

# List of Abbreviations

| | |
|---|---|
| ACCA | Automated Cloud Cover Assessment |
| ADHD | Attention Deficit Hyperactivity Disorder |
| AI | Artificial Intelligence |
| AIA | Aleut International Association |
| AMAP | Arctic Monitoring Assessment Programme |
| ASTER | Advanced Spaceborne Thermal Emission and Reflection Radiometer |
| ATR | Attenuated Total Reflection |
| ATSR | Along-Track Scanning Radiometer |
| AUV | Autonomous Underwater Vehicles |
| AVHRR | Advanced Very High-Resolution Radiometer |
| CAA | Canadian Arctic Archipelago |
| CDI | Cloud Displacement Index |
| CI | Cloud Index |
| COPD | Chronic obstructive pulmonary disease |
| COSCO | China Ocean Shipping Company |
| CSI | Cloud Shadow Index |
| DLFRF | Decision Level Fusion Random Forest |
| DSR | Dynamic Stochastic Resonance |
| EEZs | Exclusive Economic Zones |
| EGC | East Greenland Current |
| EO | Earth Observation |

| | |
|---|---|
| ESA | European Space Agency |
| ETM+ | Enhanced Thematic Mapper Plus |
| EUMETSAT | European Organisation for the Exploitation of Meteorological Satellites |
| FDI | Floating Debris Index |
| FLFRF | Feature Level Fusion Random Forest |
| FRED | Floating Robot for Eliminating Debris |
| FSI | Fog Stability Index |
| FTIS | Fourier Transform Infrared Spectroscopy |
| GBB | Great Bubble Barrier |
| GNSS | Global Navigation Satellite System |
| GNSS-RO | Global Navigation Satellite System - Radio Occultation |
| GPGP | Great Pacific Garbage Patch |
| GPS | Global Positioning Service |
| HDPE | High-density Polyethylene |
| IAC | International Astronautical Congress |
| IBD | Inflammatory Bowel Disease |
| IIPP | Institute for Innovation and Public Purpose |
| IMO | International Maritime Organization |
| IMU | Inertial Measurement Unit |
| INS | Inertial Navigation Systems |
| IR | Infrared |
| ISE | International Submarine Engineering |
| ISU | International Space University |
| IUCN | International Union for Conservation of Nature |
| JAMSTEC | Japan Agency for Marine-Earth Science and Technology |
| JAXA | Japan Aerospace and Exploration Agency |
| LCCA | Life-Cycle Costing Analysis |
| LEO | Low Earth Orbit |
| LIDAR | Light Detection and Ranging |

| | |
|---|---|
| LR | Literature Review |
| MARPOL | International Convention of the Prevention of Pollution from Ships |
| MED | Mediterranean |
| MODIS | Moderate Resolution Imaging Spectroradiometer |
| MSS | Master of Space Studies |
| NASA | National Aeronautics and Space Administration |
| NDVI | Normalised Difference Vegetation Index |
| NGOs | Non-governmental Organizations |
| NIR | Near-infrared |
| NOAA | National Oceanic and Atmospheric Administration |
| NOS | National Ocean Services |
| NSIDC | National Snow and Ice Data Center |
| NSR | Northern Sea Route |
| NWP | North West Passage |
| OCP | Ocean Cleanup Project |
| OSPAR | Oslo/Paris Convention |
| PAME | Protection of the Arctic Marine Environment |
| PET | Polyethylene Terephthalate |
| PI | Plastic Index |
| RAP | Regional Action Plan on Marine litter |
| RF | Random Forest |
| SAR | Synthetic Aperture Radar |
| SDGs | Sustainable Development Goals |
| SLSTR | Sea and Land Surface Temperature Instrument |
| SSD | Single Shot MultiBox Detector |
| SWIR | Shortwave-infrared |
| TEU | Twenty-foot Equivalent Unit |
| TIRS | Thermal Infrared Sensor |
| TM | Thematic Mapper |
| TP | Team Project |

| | |
|---|---|
| TSR | Transpolar Sea Route |
| UAV | Unmanned Aerial Vehicle |
| UCL | University College London |
| UN | United Nation |
| UNCLOS | United Nations Convention on the Law of the Sea |
| UNEP | United Nations Environmental Program |
| USD | United States Dollar |
| USF | University of South Florida |
| USGS | United States Geological Survey |
| VIR | Visible infrared |
| VNS | Visual Navigation Systems |
| WHO | World Health Organization |
| WMO | World Meteorological Organization |

# 1

# In an Ocean Not So Far Away:
# Introducing Plastic Pollution in the Arctic

*'We are tied to the ocean. And when we go back to sea, whether it is to sail or to watch it, we are going back from whence we came.'* - John F. Kennedy (1962), *35th US President.*

## 1.1 Introduction

Plastic pollution has a profound impact on all levels of the economy, impacting both land and marine ecosystems. Microplastics are a threat to human health when inhaled or ingested and can lead to financial loss for coastal businesses, fisheries, vessels and tourism (Newman, et al, 2015). There are five well documented gyres located in the various oceans around the world (5Gyres.Org, 2020). However, a sixth gyre is starting to form in the Arctic Ocean (Cózar, et al, 2017). This report will discuss the outcomes of the sixth Arctic gyre, how it can be monitored and prevented using space- based technologies in tandem with other useful airborne and waterborne technologies. Satellite services, particularly earth observation (EO) have proven to be useful for understanding the changing dynamics of plastic pollution patches and can differentiate between the different types of plastic in the oceans. Based on field observations, current estimates account for less than one per cent of total plastic pollution that enters marine ecosystems (Tekman, Krumpen and Bergmann, 2017).

This report is focused on answering the question on 'How can we monitor ocean plastic pollution using space technologies?' and submits the following mission statement:

*To propose solutions for tracking, monitoring, and mitigating ocean plastic pollution, with special emphasis on the Arctic Ocean, using space-based technologies by implementing an interdisciplinary approach to the problem in order to outline a wide-ranging policy and technical strategy as an answer to plastic pollution.*

The Arctic is particularly interesting because it is changing rapidly due to climate change and increasingly large estimations of plastic have been identified. This prompted growing concern with environmentalists, academics, indigenous populations and businesses. Floating particles that enter the Arctic Ocean from the Pacific side accumulate in the Beaufort Gyre and cross the Arctic to the Atlantic side as they travel through sea ice. There is a gap in understanding on the Arctic environment particularly on the mechanisms of sedimentation, bioavailability and degradation that take place. The processes governing plastic concentration, accumulation and transport in the Arctic are not researched enough and the Arctic is changing dramatically with the sea ice volume and thickness rapidly declining (Simmonds, 2015). These mechanisms need to be understood in order to take into account these changes in the physical environment and to estimate the evolution of plastic pollution.

There are only a few places where water and ice can leave the Arctic. The main current from the Arctic is the East Greenland Current (EGC), which flows south into the Fram Strait. The East Greenland current is also the main export mechanism for sea ice from the Arctic, carrying 4000 - 5000 $m^3$ of ice south every year (Björnsson and Pálsson, 2008). Other, smaller routes for both sea ice and water from the Arctic include either the Canadian Arctic Archipelago (CAA) or the Nares and Davis Straits.

There are two main features of sea ice drift in the Arctic Basin. Multi-year ice has survived one or more summers and is usually thicker than the first year of ice. South of the Beaufort Sea, west of the Canadian archipelago, and north of Greenland, there are known regions with older and thicker ice that survives the summer (Halsband and Herzke, 2019). The Arctic is covered by varying seasonal sea ice. It is well known that both sea ice and glacial volume have decreased over the last thirty-five years, accompanied by a thinning of sea ice in the Fram Strait (Arctic Portal, 2019). Ice thickness is crucial to

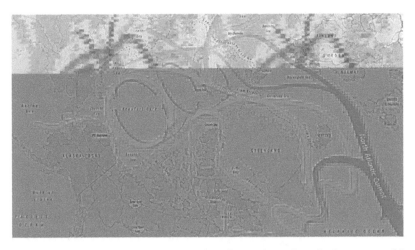

**Figure 1:** *Map of the Arctic Ocean Circulation (Woods Hole Oceanographic Institution, 2021). The Gulf Stream, brings heated Atlantic waters along the coast of Norway, flowing northward. It divides into different components, on each side of Svalbard, and travels north. This Atlantic water cools down in the Arctic Ocean, and sinks. The cold Arctic water flows across the Fram Strait between Svalbard and Greenland, after circulating in the North Polar Basin.*

study when researching the accumulation of particles because the temporary sink or transport mechanism is more exaggerated when the plastic stays longer in the ice. Figure 1 shows the Arctic Ocean circulation and member states.

High quantities of plastic objects have been reported to be trapped in Arctic sea ice because of melting sea ice; this plastic could be released into the underlying water column (Cózar, et al, 2017). However, an optimal method of monitoring is still outstanding, as beach cleaning may not reflect true changes due to large variations in weather-dependent deposition and removal of plastic litter. Pfirman, Kogeler and Anselme (1995) first introduced sea ice as a potential mechanism for the long-term transport of contaminants. The mechanism behaves as follows: frazile ice (i.e. a collection of loose ice particles) forms in supercooled water (e.g. caused by wind blowing over water at low temperatures) and scavenge particles. It is assumed that sea ice mostly grows from the ocean bottom and is efficient at capturing particles. Sea ice was found to be effective in trapping matter and sediment in ice floes in a study on Arctic ice cores (La Kanhai, et al, 2020). .

Plastic materials may also be released directly during production processes (e.g. fishing, shipping and tourism) and from local suppliers. Industrial activities and population changes along the Arctic coast are local sources of plastic pollution. Architecture, such as wastewater treatment is often absent in sparsely populated Arctic regions. There are crucial shipping routes from higher populated areas within proximity to the Arctic, which may contribute to plastic pollution but are relatively unknown to date.

Huge quantities of Atlantic water reach the Arctic Ocean through the Fram Strait, which has varying amounts of toxins and microplastics. Debris levels remained almost the same in surface waters of the North Atlantic as well as along the northeastern Atlantic coast of Ireland (Maes, et al, 2017). Microplastics embedded in sea ice for many years may be deposited in the surrounding water by melting ice, which is predicted to worsen global warming. Satellite observations of Arctic sea ice have seen a decline in magnitude for all months since 1979, shown in Figure 3 below (National Aeronautics and Space Administration (NASA) Global Climate, 2021).

The decline has coincided with the abrupt global warming of the Arctic over the last thirty years. The Arctic will function as a plastic trap as plastic builds up within these currents and continues to receive waste from mainland Europe and Scandinavia Halsband and

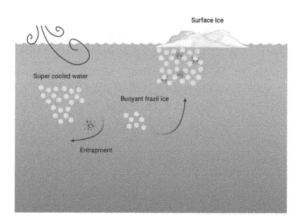

**Figure 2:** *Schematic representation of frazil ice entrapping particles. Microplastics become trapped by forming ice particles, becoming trapped within the layers of ice. When this ice melts, the plastic is released back into the ocean. Based on Figure 1 of Daly (2008)*

Herzke (2019). Relevant satellites have unprecedented coverage of the Arctic Ocean. In 2013 and 2014, there was an additional thirty-three per cent and twenty-five per cent rise compared to the 2010-2012 seasonal average (Tilling, et al, 2015). Studies on plastic detection using airborne data models and theoretical studies have shown the potential to detect macroplastics in optical data.

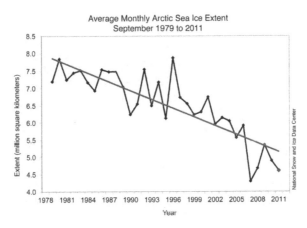

**Figure 3:** *Graph showing the melting of sea ice from 1979-2011. Peaks of higher and lower ice content is due to the change in seasons. A rapid downwards trend can be observed, which could be due to climate change (National Snow and Ice Data Center, 2011).*

The EO satellites Sentinel-2A and -2B may have improved enough (in resolution) to be of use to detect ocean plastic. The high spatial resolution of up to 10 m enables the detection of 'small' objects in the sea. The plastic marine aggregation regions may have formed according to current ocean dynamics. Plastic pollution is likely to occur in the highly contaminated North Atlantic region, as well as in the Siberian and Canadian rivers (Biermann, et al, 2020). It is clear that contaminants are prevalent in the Arctic climate, but the existing coverage of data is inadequate to map these complexities (*loc. cit.*).

The Arctic is not substantially less affected by waste produced in the higher populated areas further south. How Arctic conditions affect movement, sedimentation, bioavailability and degradation is poorly understood and not well known. Hazards to the natural environment such as climate change, emission of pollutants and ocean acidification can have synergistic or additive effects on plastic debris. Few records of in situ samples of microscopic waste in Arctic and sub-Arctic regions indicate an elevated amount of plastic present ingested by seabirds (Kühn, Bravo Rebolledo and van Franeker, 2015).

Satellite remote sensing is the world's leading technology for collecting high-quality, structured optical imagery on a global scale. However, few studies have been able to classify floating microplastics in the marine world. Data show that zooplankton at the bottom of the food chain also eats microplastic (Taipale, et al, 2019) which strongly suggests that biota in the North Arctic are at risk and could experience harmful second-order effects and, or, biomagnify plastics throughout the food chain. Specific satellites offer unprecedented coverage of the Arctic Ocean; five years of observation revealed a fourteen per cent decrease in sea ice volume between 2010 and 2012 (Beaumont, et al, 2019). However, an additional thirty-three per cent and twenty-five per cent increase was noted within 2013 and 2014 in comparison to the 2010-2012 seasonal mean. Previous studies on limiting factors have included the temporal, spatial and spectral coarseness of available data. Landsat satellites, for example, offer nine spectral bands with a spatial resolution of thirty meters and a time resolution of sixteen days. Commercial satellites, including SkySat and RapidEye, collect images at a submeter of up to five meters of spatial resolution (Biermann, et al, 2020).

According to current ocean dynamics, plastic marine aggregation regions may have formed. Plastic contamination will probably occur in the highly polluted North Atlantic region and also in the Siberian and Canadian rivers. Sea ice melting will cause an accumulation of plastic debris in these areas. Limited evidence indicates that the Arctic is not significantly less affected by waste generated within the higher populated areas further south. It is obvious that pollutants are prevalent in the Arctic environment but present data coverage is unable to map these complexities. An improved understanding of degradation processes, biota interactions, bioaccumulation, possible chemical leakage and related toxicological effects is needed. Natural disasters may also have synergistic or additive effects. Lack of research into in situ samples of microscopic waste in Arctic and sub-Arctic regions indicates elevated levels of plastic present ingested by seabirds.

## 1.2 Aims & Objectives

Our aim is to do a case study increasing understanding of plastic debris in the Arctic Ocean in order to propose a solution to prevent a sixth gyre from forming and to apply this to other oceans. To achieve this we will focus on doing a thorough review of the challenges in the Arctic Ocean and investigate existing and/or prototype technologies. These efforts will then be complimented by space sector capabilities in order to provide an answer on how to effectively monitor plastic globally with special focus on the Arctic. The table below describes the Objectives:

**Table 1:** *Aims and Objectives of the Team Project*

| N# | Description |
|---|---|
| 01 | Perform a wide-scope interdisciplinary review about the problem of plastic pollution in the oceans, including:<br>• Its sources and its consequences on society, economy and policy<br>• A review of space-based and air-based technologies for tracking and monitoring plastic in the oceans<br>• Examples of existing mitigation techniques |
| 02 | Analyze the current context of plastic pollution in the Arctic Ocean by:<br>• Describing the importance of preserving this region<br>• Discussing the outcomes of the sixth Arctic gyre on indigenous populations and economic activities<br>• Considering the best ways to coordinate an environmental preservation effort |
| 03 | Identify and analyze technologies with space or near-space components to detect, track, and mitigate plastic debris in the Arctic Ocean, by:<br>• Discussing the major challenges an ice-covered ocean can pose to monitoring<br>• Exploring cutting-edge technologies that can complement the limitations of satellites<br>• Developing an integrated strategy that can be scaled to other oceans |
| 04 | Propose a communication strategy that could raise awareness about plastic pollution in the oceans, through:<br>• A virtual symposium for governments<br>• Social media for the young generation<br>• A clean-up initiative organised by ISU |
| 05 | Recommend a series of considerations to make, along with some actions to undertake, in order to better face the problem of plastic pollution in the oceans. |

## 1.3 Methodology

This team research project adopted a mixed approach to gain a better understanding of how space can be used to monitor the ocean and track plastic. Initial stages began with a broad literature review (LR) combined with various interviews provided by the faculty interfaces to understand what initiatives organizations were undertaking to mitigate and clean up plastic pollution. Following the LR, focus was shifted from the Rhine river to the Arctic Ocean to negate the drawbacks of focusing on a river instead of an ocean. In order to ensure an interdisciplinary approach was taken, the research and analysis was divided into different subjects enabling team members with prior knowledge to undertake similar research or shift towards a new field.

To overcome the gap in literature on plastic pollution in the Arctic Ocean, further interviews were conducted to understand the difficulties surrounding the Arctic, methods for detecting plastic in ice and cloud covered conditions, obtain insights for developing an integrated strategy and information on environmental policy approaches. As an extension of the LR, research continued to determine what the extent of marine plastic pollution problem is impacting the Arctic populations and industries, Earth's climate and ocean circulation patterns. Attempts were made to see if a link exists between increased shipping activity and plastic pollution. Policy and law, business and management, space applications and humanities were investigated because of the economic entanglements of the Arctic Ocean as a sixth gyre. An in depth analysis was taken to apply space-based technologies found in the LR to the Arctic Ocean with added investigation of other technologies such as autonomous underwater vehicles and stratospheric balloons to compensate for gaps in space-based EO.

Attempts to obtain quantitative data for the costs associated with marine plastic, new Arctic trade routes, data on shipping vessels, and costs of the proposed integrated strategy was made. As this is a novel and timely study, access to information or even existence of information was limited. Use of ISU's library resources and faculty, academic journals and articles, environmental organizations and databases was used to gather and synthesise data and information. Due to the novelty and recent attraction around this topic, much of the research and data analyzed came from secondary sources and interviews conducted.

### 1.4 Report Structure

This report is structured in six chapters, each describing different aspects of the same topic, so that they are all interconnected and linked by the common thread of plastic pollution in the oceans.

### Chapter 1: In An Ocean Not So Far Away: Introducing Plastic Pollution in the Arctic

In this first section there is a brief presentation of the problem to be addressed and its effects on the ecosystem. In particular, it is discussed why the Arctic Ocean is worth this level of focus and how important it is to find a strategy to address the problem as soon as possible, before a sixth gyre forms. In addition, the potential of space technology to monitor and track plastic routes is highlighted, a tool that, combined with in situ monitoring, will be able to inform decision-making on the extent of the problem.

### Chapter 2: Full Of It: Plastic Pollution in the Oceans

Chapter 2 provides an interdisciplinary review about causes and consequences of plastic pollution, based on the LR performed by the team during a previous phase of the project. Specifically, it describes the current extent of the problem, outlining the causes and adverse effects on the marine and terrestrial ecosystem, providing the reader with all the elements necessary to build a knowledge base. This analysis demonstrates how the resources available can be used to enhance current efforts and lead to the development of applicable future technologies for remote sensing.

### Chapter 3: The Tip of the Iceberg: Plastic Pollution In the Arctic Ocean

Chapter 3 provides an analysis of the extent of the plastic pollution problem in the Arctic Ocean, identifying the causes that are leading to the formation of a sixth gyre. It also provides a survey of the effects of ocean plastic pollution on indigenous peoples, new opportunities for Arctic shipping lanes and potential increases in plastic pollution due to increased marine activities. Finally, it considers whether an Arctic Treaty is necessary or whether it is sufficient to enhance the policy tools already available to Arctic states, in order to coordinate environmental conservation efforts.

**Chapter 4: Houston, We Have A Problem: Space Technologies to Monitor Plastic in the Oceans**

Chapter 4 aims at presenting technologies with spatial or near-spatial components to detect, track, and mitigate plastic debris with state-of-the-art technology to obtain data, thus serving as a first step in preparing mitigation efforts, by assessing the advantages and limitations of such technologies. It outlines how currents in the oceans and water temperature can be monitored to study gyre formation, also describing the challenges related to the problems of reflecting and differentiating between plastic, ice and clouds to monitor plastic from space. Chapter 4 will then look at new, state-of-the-art technologies to overcome some of these limitations, thus developing an integrated strategy to apply these technologies to the Arctic Ocean.

**Chapter 5: And Action: Community Education and Investment**

Chapter 5 proposes a communication strategy that could be useful for different auditors, in particular for governments or policy-makers and for the general public - including young generations, in order to raise their awareness regarding plastic pollution in the ocean. It provides some examples of initiatives and tools that can be implemented at different levels of outreach.

**Chapter 6: Putting It All Together: Conclusions and Recommendations**

The last chapter supplies the reader with an overview of the conclusions and recommendations extrapolated from the entire project. The aim is to lead the attention of the stakeholders towards the main challenges posed by plastic pollution in the oceans, at the same time also providing them with a series of instruments potentially useful to mitigate the problem.

# 2
# Full of It: Plastic Pollution in the Oceans

*'We are living on the planet as if we have another one to go to'* -
Terry Swearingen (2007), *Nurse/Environmentalist.*

## 2.1 Introduction

It is estimated that eight million metric tons of plastic enter the oceans
every year (Topouzelis, Papakonstantinou and Garaba, 2019). Plastic
pollution has a profound impact on all trophic levels and biological
levels. Microplastics are a threat to human health when inhaled or
ingested, which will be discussed in this chapter too. Coastal cities,
businesses, fisheries, vessels and tourism are also at risk. The United
Nations (UN) set seventeen Sustainable Development Goals (SDGs)
for how society can achieve sustainable development by 2030. The
adoption of space technologies has seen ground-breaking innovations
in disaster risk management, navigation, EO and ocean temperature
monitoring, to name a few. Less has been explored in the way of
using space to track and monitor anthropogenic debris in the ocean.
Satellite technologies should be capable of detecting the presence of
different types of plastic in the oceans.

An interdisciplinary approach to 'Tracking Plastic Pollution in
Oceans from Space' study was adopted in the previous literature
review. Based on four main questions, the following subjects were
discussed: why do we use so much plastic, why is plastic contamination
a threat to life on Earth, how does pollution affect society, and what
can be done to fix the problem. The large volume of plastic in the
oceans affects the entire environment. Plastic has reached even the
most remote parts of the globe, such as the Arctic and Antarctic. The
latest satellite technology could be employed for detecting different

types of plastic in the ocean. An integrated suite of sensors is required to track and classify plastic. This analysis demonstrates how the resources available can be used to enhance current efforts and lead to the development of applicable future technologies for remote sensing.

## 2.2 Why Do We Talk About Plastic Pollution in the Oceans?

A more recent estimate that plastic debris in the ocean reached as high as fifty-one trillion microplastic pieces worldwide (Agamuthu, et al, 2019). However, there are uncertainties on the exact figure largely because it is difficult to quantify how much plastic is in the oceans due to the variety of differently-sised plastic and how much it has spread throughout the water column underneath polar ice.

### 2.2.1 The Current State of Plastic Pollution in the Oceans

Although plastic is not biodegradable, it can be fragmented and degraded by weather conditions. Depending on the type of plastic, the rate of degradation differs. For instance, high-density polyethylene (HDPE) is much more durable than others that will fragment into thousands of microplastics after a shorter time span (Ryan, et al, 2019). In Table 2 is an estimation of plastic particles likely found in oceans, according to their weight. The estimated amount of floating plastic in oceans, 0.33 mm in diameter or greater, is estimated to be 5.25 trillion particles, with a weight of 268,940 tons (Eriksen, et al, 2014). Plastic is used everywhere in modern society, and it is created for all uses from car parts to medical devices to single-use plastic utensils. Due to the advancements and innovations in plastic materials, the utilization of plastic in society has exponentially increased to the point that the twentieth century has been coined the Plastic Age (D. Lewis, 2016).

Notable milestones in plastic development include Eduard Simon (1789-1856), who invented polystyrene, then Alexander Parkes (1813-1890) with parkesine and Leo Baekeland (1863-1944) who created the very first synthetised plastic, bakelite. Following World War II, plastic consumption continued to expand and its production augmented significantly in the world as is shown in the graphic.

**Table 2:** *The estimated counts of plastic particles in the oceans in tens of billions and estimated weight of plastic in hundreds of tons (Eriksen, et al, 2014).*
*NP: North Pacific, NA: North Atlantic, SP: South Pacific, IO: Indian Ocean, MED: Mediterranean.*

|        | Size Class [mm] | NP | NA | SP | SA | IO | MED | Total |
|--------|-----------------|------|------|------|------|------|------|-------|
|        | 0.33-1.00 | 68.8 | 32.4 | 17.6 | 10.6 | 45.5 | 8.5 | 183.0 |
|        | 1.00-4.75 | 116.0 | 53.2 | 26.9 | 16.7 | 74.9 | 14.6 | 302.0 |
| Count  | 4.76-200 | 13.2 | 7.3 | 4.4 | 2.4 | 9.2 | 1.6 | 38.1 |
|        | >200 | 0.3 | 0.2 | 0.1 | 0.05 | 0.2 | 0.04 | 0.9 |
|        | **Total** | **199.0** | **93.0** | **49.1** | **29.7** | **130.0** | **24.7** | **525.0** |
|        | 0.33-1.00 | 21.0 | 10.4 | 6.5 | 3.7 | 14.6 | 14.1 | 70.4 |
|        | 1.00-4.75 | 100.0 | 42.1 | 16.9 | 11.7 | 60.1 | 53.8 | 285.0 |
| Weight | 4.76-200 | 109.0 | 45.2 | 17.8 | 12.4 | 64.6 | 57.6 | 306.0 |
|        | >200 | 734.0 | 467.0 | 169.0 | 100.0 | 452.0 | 106.0 | 2028.0 |
|        | **Total** | **964.0** | **564.7** | **210.2** | **127.8** | **591.3** | **231.5** | **2689.4** |

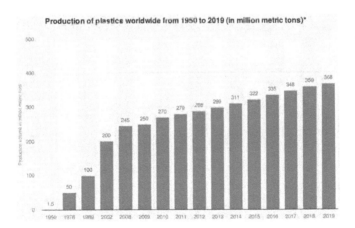

**Figure 4:** *Production of Plastics (Statista, 2020)*

Figure 4 shows that from 1950 to 1989, plastic production has multiplied by 100, due to an economic increase. Then, from 1989 to 2002, the production doubled. The plastic industry vastly expanded and it is impossible to imagine life without it now.

Plastic debris can be classified within four main categories: macroplastic, mesoplastic, microplastic and nanoplastic. Macroplastics are larger and can survive after several decades in the oceans. Plastic bottles for example belong to this category. Notably, they travel faster than other debris due to their high profile in the water. Wind can easily transport plastic bottles from point to point, quickly enough that its writing is still readable before wearing away. Around 480 billions of plastic bottles are created each year and none of the oceans in the world are safe from this pollution (Ryan, et al, 2019).

Monitoring macroplastic can encounter challenges due to the way in which biofouling organisms, such as barnacles and mussels, attach to them and cause them to become heavier, just as they would on sailing vessels, thus causing them to sink into the ocean (Lebreton, et al, 2018; Moore, et al, 2001). This phenomenon can lead to a misunderstanding of the real amount of plastic litter (Ryan, et al, 2019), as shown in an example in the Atlantic, where a 2-liter water bottle with a normal weight of forty-five grams could weigh 580 grams due to barnacle growth *(loc. cit.)*.

Ocean plastic does not only originate from land pollution but also from shipping, fishing and aquaculture which represent 28.1% of all plastic debris (Lebreton, et al, 2018). Marine shipping accidents can also cause large inputs of pollution, as in the case of the sinking of the MOL Comfort that caused more than 4,000 containers and their merchandise to be discharged into the oceans (World Shipping Council, 2017). Mesoplastic is a kind of macroplastic classified between 5 mm and 2 cm (Duncan, et al, 2018), while microplastics, with a diameter of 5 mm or smaller, represent 92.4% of marine plastic particles and are more difficult to detect (Eriksen, et al, 2014). As detection methods dedicated for microplastic increase, the amount of total mass identified in the ocean will significantly increase as well *(loc. cit.)*.

Nanoplastics are plastic particles with a size of less than 100 nm, which are the products of the decomposition of microplastics. The World Health Organization (WHO) team started to examine the risks associated with microplastic pollution in bottled water in 2018. It is not shocking that bottled water contains hundreds of nanoplastic particles. After two years of testing, the WHO still does not have clear evidence of the health effects of plastic on daily drinking water. On the other hand, another recent study using electron scanning microscopy has shown that the food-grade nylon and polyethylene terephthalate (PET) used in tea bags is a significant source of micro and nano plastic ingested by humans. The mechanism causing this supposedly food-grade material to decompose so rapidly is not yet effectively understood (Hernandez, et al, 2019).

### 2.2.2 Sources of Marine Plastic Litter

When plastic enters the ocean, it follows different routes, circulating via ocean currents. Plastic can be expelled from ships, shorelines or land and guided into various drifts and directions (see Figure 5) (Kane, et al, 2020). Due to the combination of meteorological conditions (weather and wind flows), currents have generated gyres of garbage patches, representing a wide zone of contamination (NationalOceanic and Atmospheric Administration (NOAA), 2013).

These gyres are named North Pacific Gyre, North Atlantic Gyre, South Pacific Gyre, South Atlantic Gyre, and Indian Ocean Gyre (Figure 6). Within these Gyres are five strong currents responsible for the hot and cold temperature distribution, marine nutrients and salinity (NOAA, 2020). They are named the Western Boundary Current, Eastern Boundary Current and two Transverse Currents, which include whirlpools bringing a rotative movement to the Gyres that captures the debris (Evers and Editing, 2019).

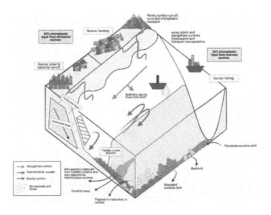

**Figure 5:** *Ocean floor currents direct microplastics. Visual representation of different currents in the transport, accumulation and distribution of microplastic in the deep ocean. Along the shelf currents microplastics disperse. High-gravity currents direct microplastic into the ocean floor. Thermohaline currents segregate microplastic into localised concentrated patches. Adapted from (Kane, et al, 2020).*

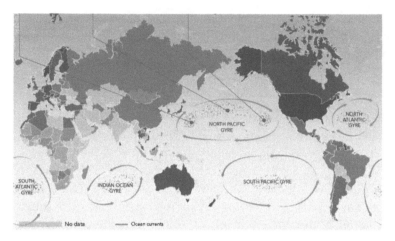

**Figure 6:**  *5 Ocean Gyres, Plastic Accumulation (Evers and Editing, 2019).*

The most prominent of the five Gyres is named the Great Pacific Garbage Patch (GPGP) (see Figure 7). It is composed of three parts, one coming from the Japanese coast, the second part is emitted from the Californian coast, and in the middle, is the Subtropical Convergence Zone. This phenomena is due to the role played by warm water from the South Pacific and the cold water from the Arctic (Evers and Editing, 2019).

In addition to being the largest and most polluted gyre, the GPGP seems to be most dense relative to its radius of 1,100 km and contain more than a million pieces of debris per *km²* (Duncan, et al, 2018). The NOAA Marine Debris Program estimated that sixty-seven ships working for one year to clean the place would only clean up one per cent of the total pollution (Evers and Editing, 2019; Law, et al, 2014; Lebreton, et al, 2018).

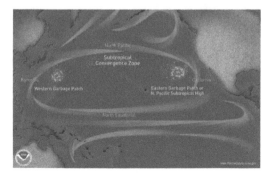

**Figure 7:**  *Great Pacific Garbage Patch (Gibbons, 2019)*

**North and South Atlantic Gyre**
It is estimated that the North Atlantic Ocean, innervated by strong currents, nevertheless contains less than half of the plastic debris present in the North Pacific at 930 billion pieces but nonetheless remains more than triple the estimated values for the South Atlantic (Eriksen, et al, 2014). It should be noted that eighty-three per cent of the plastic found in the North Atlantic gyre is located between 22°N and 30°N while the South Atlantic has less plastic than in the northern half of the same ocean (Law, et al, 2010). Estimates suggest that 297 billion pieces weighing 12,780 metric tons (Eriksen, et al, 2014) float in the South Atlantic. Wind and waves seem to drive most debris from coastal areas towards the gyre (Law, et al, 2010).

**Indian Ocean Gyre**
In the Indian Ocean, plastic debris exceeds the rest of the Southern Hemisphere by about 591,300 tons (Eriksen, et al, 2014). Torrential rains and a large number of people living near the Indian Ocean are assumed to carry plastic into the ocean. The Indian Ocean Gyre is of interest as a result of its monsoon system, which effectively pushes plastic pollution further west into the Pacific whorl (Maes, et al, 2018); this phenomenon is narrowing in the meantime. With its particular geography, currents and atmospheric conditions, there is an east-west countercurrent from the equator transporting submerged plastics to Australia, where monsoon winds spread plastic particles to its coast (Riskas, 2019).

**Mediterranean Sea**
The Mediterranean Sea with more than 17,000 marine species (Coll, et al, 2010) is a biodiversity point rich in wildlife, marine life and coastal tourism. The World Wildlife Fund warns the countries of the Mediterranean that a 'sea of plastic' may be due to tourism in Turkey, Spain, Italy, France and Egypt (Naware, 2018), which also poses a risk to marine life in the area. Furthermore, this pollution has had an impact on regional air quality. Indeed, a study proved that microplastics in Paris were present in the atmosphere and averaged a fallout of 118 particles per square meter per day (Dris, et al, 2015). Furthermore, the Mediterranean has one of the highest densities of microplastics of around 250 to 274 billion total fragments or 109,000 particles per square kilometer (Abreu and Pedrotti, 2019; Duncan, et al, 2018; Eriksen, et al, 2014), with plastics largely flowing in from the eastern half of the sea.

## Polar Regions
In the Arctic Ocean, located between East Greenland and the Svalbard Islands is the Fram Strait. It has the peculiarity of containing one of the highest concentrations of microplastics worldwide (Katz, 2019). More than 12,000 microplastics per liter are present in the ice of the coastal towns of the Arctic Circle *(loc. cit.)*. This high concentration of plastic in the north led to the creation of a sixth garbage vortex in the Barents Sea just above Norway and Russia. Oceanographer Erik van Sebille (Utrecht) explains that this is probably due to the fact that the waters of the Atlantic in the north direction are cooling and therefore sinking, creating the turning circulation of the Atlantic meridian *(loc. cit.)*. In addition, due to underground currents, plastics around the Antarctic coast have become trapped in the ice (Doyle, 2018).

## Estimated Quantity of Plastic Pollution
Experts state that almost eighty per cent of the Earth's plastic pollution arrives from land use and around twenty per cent would come from offshore platforms. Plastic waste coming from household garbage cans, industrial waste, vehicles and agriculture seems to catalyze plastic ending up in rivers and streams, and subsequently carrying toxic matter into the ocean (see Figure 8) (Horton, et al, 2017).

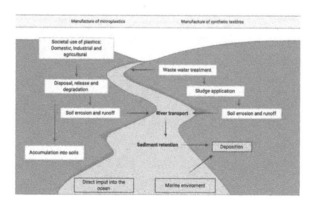

**Figure 8:** *Plastic waste manufacturing route and disposal from major polluting rivers and at the sea surface. Adapted from (Horton, et al, 2017).*

A huge quantity of plastic is found on beaches or coastlines and similarly ends up in the sea. All of this will further increase the amount of 412 million tons of plastic per year that flows into the oceans. Additionally, the world's population is constantly growing and currently sits at roughly 7.7 billion inhabitants (U.S. Census, 2021).

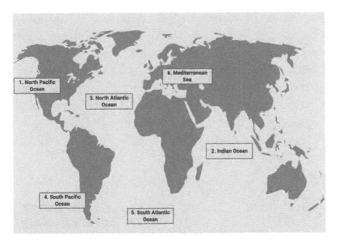

**Figure 9:** *Map of world oceans: Six main oceans ranked by highest amount of plastic pollution. 1. North Pacific 2 trillion. 2. Indian Ocean 1.3 trillion. 3. North Atlantic 930 billion. 4. South Pacific 491 billion. 5. South Atlantic 297 billion. 6. Mediterranean Sea 46.3 billion. Adapted from (Eriksen, et al, 2014).*

## 2.3  We Are What We Eat: Effects of Plastics On the Food Chain

There is a lack of information on the precise impact of plastic waste on the aquatic biosphere due to the difficulties of observing biodiversity in nature. The ever-increasing volume of plastic in the oceans is the reason why it is more important than the naturally occurring floating waste. Entanglement, ingestion and chemical degradation are the major forms of plastic pollution affecting marine wildlife and are at the greatest risk of ingestion (Beaumont, et al, 2019). Plastic contamination affects the entire food chain of marine ecology. Three of the main taxa threatened by macroplastic contamination are seabirds, sea turtles and marine mammals. Plastic pollution kills vegetation, can cause direct mortality of seabird eggs and eliminates natural food sources. The harmful impact of particles with diameters less than 5 mm (microplastics) and less than 100 nm (nanoplastics) has raised serious concerns worldwide (Guo and Wang, 2019). They are often eaten by marine animals, who confuse them with food, even in the most remote areas of the ocean. This may harm the early stages of life of marine organisms and may hinder the reproductive patterns of marine organisms. For example, plastic ingestion may alter the role of the endocrine system in fish (Rochman, et al, 2014). Plastic particles have been found in various animals from the lowest levels of the food chain, such as zooplanktonic organisms, to the highest levels of both invertebrates (crustaceans, molluscs) and vertebrates (fish). A study found that after obtaining 1,822 microplastic particles from the stomach and intestines of 1,337 fish specimens along the Mediterranean coast, the majority of the ingested particles were fibres (seventy per cent) and hard plastics (20.8%) (Güven, et al, 2017).

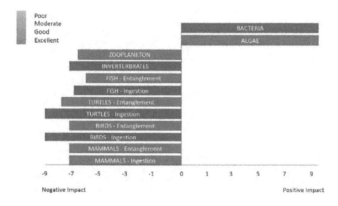

**Figure 10:** *Impact of marine plastics on biota in the ecosystem. A score of 9 indicates global, highly irreversible, and high frequency lethal or sub-lethal effect; a score of +9 indicates: global, highly irreversible, and high frequency positive effect in terms of diversity and/or abundance (Beaumont, et al, 2019).*

Floating aquatic debris has been associated with invasive species distributing in offshore, intertidal or coastal habitats. Plastic is a vector for non-native organisms over long distances due to the vast volume of floating debris that it is non-biodegradable. The ever-increasing amount of plastic in the oceans is the reason why it is more important than naturally occurring floating debris such as volcanic rocks, wood and algae (Lewis, Riddle, and Smith, 2005).

### 2.3.1 Effects on Human Health

A poignant problem is the significant body of evidence that plastic is not only in the seas but also in marine life and ultimately the humans that eat it. Human health adverse effects remain a subject of intense debate, although there is a growing agreement that plastic and their additives are not as harmless as they were once thought to be. Cellulose and microplastics have also been found in non-neoplastic and cancerous lung tissues originating from patients with multiple types of lung cancer (Prata, 2018). These particles have shown no degradation, supporting the possibility that they would be bio-persistent. These results suggest that the human trachea is wide enough to allow plastic fibers to enter the deep lung. Microplastics 135 μm in length were estimated to be approximately one-fourth of the seventeen-generation respiratory bronchiole diameter (540 pm in diameter, 1,410 pm in length) (Awuchi and Awuchi, 2019). These results indicate that some fibers are resistant to clearing processes and can, as they proceed, cause acute or chronic inflammation, which may lead to other potential health problems, such as gastritis, chronic obstructive pulmonary disease (COPD), inflammatory bowel disease (IBD) and more.

Research needs to be done on how biotechnological plastic can be 'filtered' from marine food sources. The important questions that need to be raised are as follows: should plastic-derived toxicology indicators be used in general health evaluations; and are there ways in which the amount of potentially harmful chemicals present in plastic waste can be minimised? Unclean beaches have a higher bacterial abundance, which can contribute to higher rates of disease transmission (Malm et al, 2004). Additionally, floating marine waste helps invasive species to spread. Seafood demand has doubled over the last fifty years and the global population is projected to rise to ten billion by 2050 (Kaye and Warner, 2018).

### 2.3.2 Effects and Perspectives of Society, Economy, and Policy Society

Literature is limited in regard to the ways in which plastic pollution influences society and vice versa. This may be, in part, due to the difficult nature of quantifying such an endeavor. Indeed, characterizing the impact of plastic pollution on society in respect to local, national and international scales is challenging.

To begin, social perception and social science research regarding plastic pollution shall be discussed. 'The Blue Planet Effect' coined in the aftermath of the film 'Blue Planet' which aired in 2001 refers to society's increased awareness of ocean plastic pollution as a result of understanding its current foothold and influence in the modern world (Henderson, 2019). However, this was not the only documentary to 'hit home' to establishing a societal conception of ocean plastic pollution. A documentary produced by Leeson (2016) titled 'A Plastic Ocean' focused on single-use plastics, such as plastic bottles, and the detrimental impacts of such plastics on society. People are not very aware of the impacts of ocean plastic pollution on their society, which was affirmed by Henderson (2019), who stated that, 'social scientists emphasised that we did not yet know enough about human actions, media reporting or social practices concerning the topic [of ocean plastic pollution]'. As stated earlier, much is to be gained in social scientific research regarding the topic. Though, there are some notable studies, which shall be explained below.

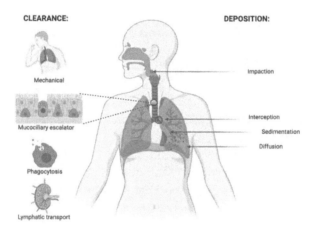

**Figure 11:** *Visualization of natural lung microparticle expulsion and prevention vs paths of microplastic inhalation and ingestion (Prata, 2018).*

Pahl and Wyles (2017) made an attempt to contribute to social science research regarding ocean plastic pollution when they discussed the 'human perception and behavior' perspective. The authors recommended a deeper understanding of norms and values as they relate to social behavior (*loc. cit.*).

In two other studies, Lozoya, et al. (2016) and Brouwer, et al. (2017) both affirm the importance of beach studies including social research about ocean plastic pollution. Lozoya, et al. (2016) goes further stating that the beach is a 'social-ecological system that provide(s) several services improving human well-being.' The beach as a literal social site should indeed be studied further as there are not enough studies on social sources of pollution (Brouwer, et al, 2017). Sufficient to say that the field is wide open for social scientists to inquire about the topic of ocean plastic pollution.

There are a few other studies of importance to mention in regard to the social impacts of plastic pollution, as well as community perception of it. Beach communities are impacted by plastic pollution. Notably, Ashbullby, et al. (2013) tackled the impact of plastic pollution on the family, specifically those living near beaches in England, highlighting the importance of beaches for social interactions. Not only does beach plastic litter decrease tourism on beaches but it also interferes with coastal life (e.g. families) (Brouwer, et al, 2017). Santos, et al. (2005) contributed the notion that one's level of education and income influence littering behavior. The work of Eastman, et al. (2013) bolstered and expanded this claim, as they found that gender, age and place of residence should be considered when considering littering behavior.

Despite the thin body of work done on the social perspectives and attitudes towards ocean plastic pollution, there is hope on the horizon. Deontologically, more people seem to be 'getting involved' in efforts to mitigate pollution. There are an increasing number of volunteers for beach clean-ups; Bramston, Pretty and Zammit (2011) affirm this, stating that volunteering encourages social belonging, fun and environmentally friendly practices.

**Economy**

As it stands, those most involved in determining the costs of ocean plastic pollution specialise in economics germane to ecology, the environment, development and resource utilization. It is difficult to properly characterise the cost of ocean plastic pollution but a study conducted by the Marine Pollution Bulletin determined global plastic pollution costs society $2.5 trillion USD annually (Szczepanski, 2019). This value was calculated per consideration of several variables including costs of air pollution during manufacturing, incineration, disposal and revenues lost by fisheries and coastal businesses. Another calculation of the cost of ocean plastic pollution was proposed by Deloitte (2019), developed a classification system that considered the qualitative, indirect impacts on society (e.g. public health, real estate, marine ecosystems) and the quantitative, direct impacts of it (e.g. clean-up costs from major pollution on the coast). In total, a conservative amount of $5.3-$14 billion USD is lost every year to ocean plastic cleanup costs in Asia from river plastic pollution alone (*loc. cit.*). This result is important because oceans and rivers are interlinked in regard to plastic pollution. Ocean plastic contamination is, to say the least, a costly problem. Currently, in order to explain the phenomenon better, it would be useful to define the types of economic structures that inform it.

When discussing how the planet has become so vast in ocean plastic waste, it is important to recognise that many businesses (plastic production or sale) are cynical of the economic advantages and disadvantages of addressing such pollution. Cleaning up one per cent of ocean plastics per year is estimated to cost between $122-$489 million USD per year, non-including equipment and labor costs (Morishige, 2012). This makes it clear that the economic advantages of manufacturing plastics may be greater than the costs associated with cleaning them up.

On the other hand, flawed chemical manufacturing paradigms have exacerbated marine plastic pollution. Such environmental accounting models only take into account three phases of the manufacturing cycle of plastics: the natural resources used, the production of plastics and the transport to landfills for disposal (Ellen Macarthur Foundation, 2017). There is only limited consideration of recycling, reducing production, and reusability of plastic (*loc. cit.*). This leads to another aspect which informs ocean plastic pollution, that is, product life cycle analysis models. An example of this is the commonly used method of pricing, called Environmental Full-Costing, or Life-Cycle Costing Analysis (LCCA). It accounts for the full life cycle of an item, from extraction of resources to disposal (Newman, et al, 2015). However, this approach ultimately fails by not accounting for the costs to society. In this way plastic pollution costs are not fully accounted for.

There are three types of costs to society: the cleanup cost (e.g. beach clean-ups, fish gear loss, etc), revenue loss (e.g. for coastal communities, businesses, and fisheries), and welfare costs relating to the lost of cultural value, aesthetic preservation, and health impacts (Newman, et al, 2015). Additionally, leakage, or plastic pollution that

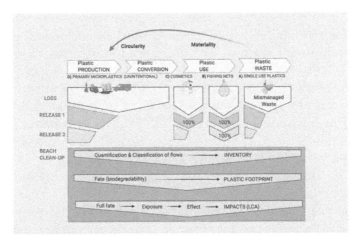

**Figure 12:** *Materiality Timeline and Leakage. Adapted from (Boucher and Billard, 2020).*

occurs at multiple stages in the plastic lifecycle, is often neglected (Boucher and Billard, 2020). Refer to Figure 12 to understand the phases that contribute to this dissipation. The aforementioned paragraphs exemplify that the ways in which society thinks about plastic production influences whether it considers the impact of ocean plastic pollution to begin with. Now that examples of cost and approaches of ocean plastic pollution production have been characterised, it is now appropriate to identify the markets which are impacted by it.

The main markets which are adversely impacted include fishing and aquaculture, marine and coastal tourism and shipping and ocean logistics. Regarding fishing and aquaculture, ocean plastics can decrease revenues from diminished catch and damage to equipment. For example, in the United States, $250 million USD in potential revenue is lost due to unsellable lobster catch compromised by pollution (IMO, 2017).

Regarding marine and coastal tourism, ocean plastic impacts beaches and their local communities. Dirty beaches do not attract tourists, which are important for supporting coastal communities. Beach litter, which is increasing at an alarming rate, deters holiday makers and may increase 250 fold in the more polluted areas in the coming decade (Eriksen, et al, 2014).

Beach debris adversely impacts tourism contributing to the need for mechanical beach cleaning. Most plastic monitoring is achieved through beach surveys of stranded litter, though sampling at sea is difficult due to high costs and the need for a larger sample size. Additionally, the Asia-Pacific Economic Cooperation region experienced less tourism revenue (i.e. decrease in $622M USD) due to marine litter (Abalansa, et al, 2020). In respect to shipping and logistics, there are costs associated with ocean plastic pollution. For example, vessel rescue costs equalled $2.8 million USD in 2008 due to the 286 rescues that were called for of which plastic pollution was to blame (Niaounakis, 2017).

Evidence indicates that marine plastic pollution will rise over the coming years and if current pollution patterns are not mitigated, by 2060 there could be 265 tons of ocean plastic pollution, resulting in significant consequences (Bergmann, et al, 2019). To highlight what was just discussed, the problem of marine plastic pollution has been well characterised in terms of its trajectory.

**Policy**
This report would be remiss if it did not touch upon the integral role that policy plays within the context of ocean plastic pollution and its mitigation. At the fundamental level, disposing plastic waste in the ocean has been illegal since the International Convention of the Prevention of Pollution from Ships (MARPOL) held in 1989 (International Maritime Organization, 1989). However, due to inadequate accountability, plastic dumping still occurs at concerning rates despite many studies' findings which have called for a decrease in ocean plastic pollution (Ryan, et al, 2019). To understand the policy- context further, an international lens will be provided below.

On the international scale, the United Nations has taken initiative to combat pollution since the 1970s, especially in regard to oceans (IMO, 2019). This was carried out at the London Convention. This was further bolstered by a subsidiary of the UN, The International Maritime Organization (IMO), which produced a treaty for preventing ships from throwing garbage into the ocean, which has been signed and ratified by more than seventy-five governments (International Maritime Organization (IMO), 1988). Efforts to reduce plastic pollution were also put forth during the International Marine Debris Conference in 2011, where a novel strategy was created and termed The Honolulu Strategy. Signatory states were encouraged to involve themselves in collaborative initiatives aimed at reducing the impact of marine debris upon health, the economy and ecological systems. Clearly, there has been global awareness of the need to address ocean plastic pollution. This notion is substantiated by the UN incorporating ocean plastic reduction and mitigation into its SDGs as the 14th priority, Life Below Water. However, it is useful to consider regional and national policy on the subject as well.

On a regional and national scale, ocean plastic pollution mitigation and reduction is embodied by a number of policies. These are primarily comprised of what has been termed as 'binding directives' and 'non-binding' action plans, where the former refers to legally binding agreements, and the latter referring to 'soft' non-legally binding agreements; such agreements follow suit with the policies outline on the international scale, and there has indeed been an increase in regional policies germane to ocean plastic pollution in the last decade (Karasik, et al, 2020). National policies regarding plastic pollution have increased in the past decade, too (*loc. cit.*). Such policy uses national law as the regulatory tool by which to enforce its agenda and is mainly used to establish acceptable levels of pollution (*loc. cit.*). This is particularly true in developed nations, where the entire supply chain is regulated, ranging from the production of plastic to its disposal (Alpizar, et al, 2020). Other examples of national policy include the complete ban of plastic bags by countries such as China, Mauritius, Kenya and Rwanda, where selling a plastic bag is a crime (Dauvergne, 2018). This action has indeed been taken on the national, regional and international levels regarding ocean plastic pollution. This now begs the question: What is being used to detect, track, mitigate and clean up such pollution?

## 2.4  What Technologies Do We Have to Solve the Problem?

### 2.4.1 Monitoring Macroplastics Through Current Space-borne Technologies

Currently, plastic sources and routes are not very well known. This section attempts to identify pollution routes and which technologies and sampling methods can track them most efficiently. The source of plastic in the water comes from human activities. Around eighty per cent of marine plastics come from land (Li, Tse and Fok, 2016).

Therefore, it is imperative to track the source and then try to mitigate its pollution. The majority of waste management is not an ideal solution, whether in the production process or after being consumed (Nielsen, et al, 2020). Plastics are also particularly menacing in comparison to other marine debris.

### Effects of Photodegradation on Plastics

One of the biggest problems with plastic pollutants and what differentiates them from biodegradable pollutants is their so-called photodegradation effect, i.e, plastic disintegrating under ultraviolet radiation. This leads to the creation of smaller plastic particles, which continue to deteriorate, into small molecules (e.g. microplastics). With the current status of technology, these particles are tough to track. To achieve this, Raman spectroscopy and luminescence analysis are used. The problem with these approaches is that these methods should be applied in-situ, which requires extensive human labor and technological resources. In order to narrow the distance and yield more positive and constructive results, this paper will only deal with macroplastics. As is the case, methods such as Raman spectroscopy and luminescence methods will not be detailed in this report as they are effective methods for the manual tracking of microplastics.

The methods mentioned above are not generally utilised to track macroplastics. The reasoning behind this lies in their fundamental design, Raman spectroscopy is exceedingly sensitive to interference, the luminescence analysis is not possible from a satellite, so it cannot be applied as EO technology either. There currently is not a reliable way to analyze microplastics pollutants remotely.

From space, it is much more feasible to focus on the treatment of macroplastics. However, a modification of the Fourier-transform infrared spectroscopy (FTIS) system may be used for remote sensing. The section below will explain the fundamentals of FTIS and how it is used to detect microplastics and how Biermann, et al. (2020) have advanced this method of creating a more efficient macroplastic monitoring device.

### Fourier-transform Infrared Spectroscopy (FTIS)

FTIS is based on the property of chemical compositions to absorb or emit light in specific conditions in the field of infrared radiation. The classical approach exposes the sample by monochromatic light, measuring the intensity of the reflected light. The light source frequency is changed, hence creating a picture of the absorption spectrum of the specific compound. This technique requires a very stable source of light, which can easily reconfigure its wavelength. In practice, such a device is hard to implement, an augmented approach of this method uses a broad light source, which contains different frequencies, to create a spectrum of absorbance. The software then processes the received data to analyze the results for a specific pattern. There are three methods of FTIS.

**Transmission Method:** Used for thin samples (<10 µm). The radiation is going through the sample and is measured on the opposite side. During this, the absorption of the sample is measured. The method is usually applicable to solid samples.

**Reflection Method:** The reflection method (Figure 13 illustrates specular and diffuse reflection methods) is usually used on samples with high reflectance properties. Based on a reflectance method lies the principle that the specific refractive index (n) for any chemical compound is changed according to the wavelength of light. As the beam of a specific wavelength falls to a sample, the refracted index's value is changed. By measuring its value, it is possible to estimate the sample's adsorption properties (Bradley, 2021).

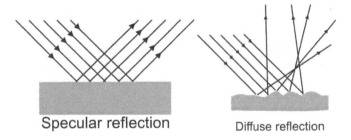

Specular reflection　　　Diffuse reflection

**Figure 13:** *Specular and diffuse reflection methods (Khoshhesab, 2012)*

**Attenuated Total Reflectance:** A conference paper by Bentley (2021), describes that it is possible to detect microplastics using near-infrared reflectance when there is a high concentration of plastic in the ocean by considering an elevation of local reflection of specific wavelengths. Observing a spectrum between 748 nm and 869 nm allows for the detection of microplastics.

In this method (Figure 14), the sample is held around a crystal made of Ge, ZnSe, or Si, which provides high optical density. This crystal is known as Internal Reflection Element, or Attenuated Total Reflection (ATR) crystal (Bell, Nel and Stuart, 2019). By design, the source of light falls to the sample under a specific angle. Due to one or multiple reflections from the sample, the energy is dissipated in it as the so-known evanescent wave (Milosevic, 2004). During this process, the sample absorbs part of the energy, which corresponds to the specific wavelength. In the end, the exit beam is analyzed and its spectrum is measured (*loc. cit.*).

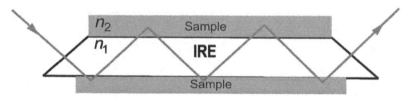

**Figure 14:** *ATR method principle. (Khoshhesab, 2012)*

In a remarkable case that makes FTIS relevant for tracking and monitoring macroplastics, Biermann, et al. (2020), introduced an approach that makes possible not only plastic tracking but also its differentiation from natural objects, such as seafoam or seaweed. The authors introduced two metrics related to each other, Normalised Difference Vegetation Index (NDVI) and Floating Debris Index (FDI), which were applied to the Sentinel-2 received data. In this study, machine learning was also used. The crucial finding of this research was that the authors managed to find specific clustering patterns of different kinds of sea garbage. This is detailed in section 4.5.1 of this report.

**Using Reflectance Spectroscopy for Macroplastic Detection and Tracking**
Reflectance spectroscopy often used in remote sensing can be used to detect and classify different materials (Garaba, Arias, et al, 2021) by comparing the difference in optical reflectance properties on plastic that has been in the ocean and plastic that has not, also referred to as virgin plastic. Ocean-harvested plastic does not have the same hyperspectral properties as virgin plastic. For instance, collected

plastic seems more isotropic. Garaba, Arias, et al. (2021), Goddijn-Murphy, et al. (2018), and Masoumi, Safavi and Khani (2012), studied the reflectance spectrum of different plastics using wavelengths in the visible, near-infrared (NIR) and short-wave infrared (SWIR) spectra. They found that the type of plastic being studied impacts the reflectance spectrum and, more importantly, depends on the angle at which the observation is taken. This means that the concentration of plastic objects and the type of water will change different plastic objects' illumination, making it possible to use direct sunlight to observe plastics in the oceans as all plastics produce detectable signals in the previously mentioned spectra.

Several concepts are presented to detect plastic in practice via reflectance spectroscopy. Garaba, Aitken, et al. (2018) demonstrate using an aeroplane that works using the SWIR and visible spectra and can detect buoyant plastic within 0.5m of the water's surface with high accuracy. Goddijn-Murphy, et al. (2018) present a concept for an algorithm to detect marine macroplastics using remote sensing data. According to Garaba, Arias, et al. (2021), a signal arriving at a satellite in lower earth orbit (LEO) above the atmosphere would suffer from too much interference due to the light's diffusion in the atmosphere. Therefore it would be too difficult to distinguish different types of plastic but still possible to differentiate them from natural objects. This statement is reiterated by Masoumi, Safavi and Khani (2012) who say that reflectance spectroscopy is not practical for plastic classification from space.

There are critical limitations to the use of reflectance spectroscopy to track and detect plastics in the oceans. The first one is that most plastics are at the bottom of the oceans, around ninety-five per cent of the total amount of plastic (Michel, et al, 2020). The second one is tied to the first one, as electromagnetic waves used for remote sensing are mostly blocked by this amount of water. Thus it is only possible to use reflectance spectroscopy for plastic on the surface or at shallow depth (Ouadih, Ouladsaid and Hassani, 2011). Therefore, most of the plastic is not going to be directly detectable. Tracking floating plastic could help develop models to identify where plastic debris originates and where it sinks.

Another problem is that EO techniques will collect enormous volumes of raw data. The employment of artificial intelligence (AI) and deep machine learning will be needed for processing the data in meaningful ways (Michel, et al, 2020). Finally, according to Michel, et al. (2020), satellites are not the most suitable technology to track plastic in rivers. Other methods will be needed to detect and track river pollution.

It is essential to consider when and why reflectance spectroscopy should be used to detect and track plastics in the oceans. First, remote sensing satellites require illumination of all objects being targeted for tracking. This illumination can be either active or passive for reflectance information. Active remote sensing could be a better option to ensure a constant stream of new data and adequate coverage. It is possible to detect plastics quite easily with reflectance spectroscopy. However, it will prove challenging to identify the type of plastic from a satellite (Masoumi, Safavi and Khani, 2012).

It is possible that a constellation of satellites for tracking and detecting plastic would need to be complemented by other means of detection such as drones, planes and measurements in the water to provide sufficient data. This enormous amount of data would then necessitate the use of deep learning or AI to process it and establish predictions on plastics' movement in the oceans, ergo bypassing the use of slow and complex models.

### 2.4.2 Other Technologies for Tracking and Monitoring Macroplastic Pollution

#### Airborne Technologies

With the rise of the remote sensing industry, it is now possible to track and monitor microplastics in oceans using satellite optical data. There are two types of plastics: those that sink and those that float on the ocean surface. Plastics in the sea came from two sources: terrestrial sources that account for between 4.8 and 12.7 million tons of plastics annually and river sources representing around two million tons annually (Martínez-Vicente, et al, 2019). Through remote sensing, spectroscopy can detect attributes of plastic absorbance peaks and differentiate it from the seaweed and other sources absorbing molecules, e.g, as previously mentioned, ESA's Sentinel-2 satellite (Biermann, et al, 2020) can detect plastics in coastal water and the deep ocean. The more plastic analyzed, the wider variety becomes detectable and the easier sources can be mitigated.

Coastal countries where plastic is a problem should invest in plastic characterization via remote sensing platforms. ESA provides free access to remote sensing satellites. More recently, thanks to the Copernicus program, a model able to track particles in different water layers have emerged (Copernicus, 2021). With time and more data, plastic detecting algorithms' evolution might become vital to cutting off plastic routes.

Using spectroscopy, most plastics display a significant optical density around 850 nm but given the diversity of plastic debris, no common signature in spectroscopy was found for those plastics (Serranti, et al, 2010). Another airborne technology used for marine plastic is the 'CLEVER-Volume' system that aims to reduce marine litter in the ocean. This technology uses Light Detection and Ranging (LIDAR) technology for a geometric waste assessment system that will extrapolate waste volume, thus knowing if any 'loss' of waste occurred.

It is also possible to track plastic using active trackers such as a global positioning system (GPS) tracker to obtain a detailed route that it and other debris have taken (Emmerik and Schwarz, 2020). A similar study has also been conducted through the Seine river with GPS trackers (Tramoy, et al, 2019). This study was useful to assess plastic streams in the Seine. The results could set a precedent to create a common and universal standard of plastic debris in waterways globally.

Plastic can be monitored in oceans through satellites and sensors, and currently, Sentinel-2 satellites are used by most researchers for monitoring plastics. Table 11 in Appendix A lists a selection of different satellites used for monitoring the oceans.

**Waterborne Technologies and Combining Existing Technologies**
A prominent source of plastic entering the ocean is rivers (Law, 2017). If the routes can be efficiently tracked, methods to mitigate the threat of ocean debris entering the sea can be established. Tracking plastic pollution routes is a relatively new science. Still, studies show eighty-six per cent of plastic entering the ocean from rivers comes from countries such as China, India, Indonesia and Bangladesh (Lebreton, et al, 2017). The majority of mismanaged waste comes from the Philippines, Vietnam and Sri Lanka (Jambeck, et al, 2015). This indicates where to focus current technologies for detection,

tracking and mitigation. Plastic concentration in rivers is directly proportional to the nearby population. In Europe, population density near rivers is higher than in Africa. For coastal areas, marine debris education and awareness is needed.

Methods that can be used as a foundation to advance studies in tracking pollution routes are modelling approaches which can be focused on areas of high pollution or ideally every river in the future (*loc. cit.*). Another tracking method uses in situ plastic cluster sampling around existing infrastructure called passive sampling (Gasperi, et al, 2014). Capturing the plastic from rivers before it enters the ocean needs to be progressed as presently only a few rivers have this capability.

It would also allow sampling from the sediments, sea or river floor and marine debris. Liedermann, et al. (2018) describe this sampling method's advantages in its flexibility, distribution, regularity and larger sample size. This makes the study more complete and allows thorough testing of effects such as meteorological conditions. The difficulties in implementing this in comparison to passive sampling require an increase in assets, supplies and funding.

The objective of using technology in the fight against ocean pollution is to provide an accurate and complete model of the waste in the oceans globally. This is a key objective as it will allow targeted clean-up actions and, more importantly, real-time information about plastic waste behaviour. While the large number of technologies that have been explored provides viable ways to track the plastic in the oceans, it remains difficult to accurately track waste routes in the oceans. A well-defined model has not yet been established. The complexity of establishing such a model is apparent when considering factors such as the origin, vast movement, marine life interaction and physical evolution of debris in the ocean. Other factors are more predictable, such as ocean currents. Several models of ocean currents have been proposed over the last few decades, ranging in complexity (Omerdic and Toal, 2007). The physical properties responsible for currents are well understood and documented, rendering their modelling a feasible task.

How can one produce a complete model of plastic movement in the ocean if a large unquantifiable quantity is introduced in the system from unknown locations around the globe every day? (International Union for Conservation of Nature (IUCN), 2020). According to Doctor Su-Yin Tan from Waterloo University, the answer lies in combining multiple technologies and approaches providing data in realtime (Tan, 2020).

The technologies described all have exciting applications and provide powerful means to learn plastic's behaviour in the ocean. All these offer a method of tracking a specific phenomenon or group of plastics. None of the methods discussed allow for a comprehensive study of the subject on its own. For example, satellite and airborne technologies have been successfully used to locate large pieces of plastic over extended areas of the ocean (Biermann, et al, 2020). Infrared and other non-optical imagery also allows information to be gathered at night and during cloudy, rainy weather making it a pivotal contribution to continuous tracking. However, satellite and airborne technologies are limited to tracking plastics in the order of several centimeters or more, making them an essential yet incomplete solution. The reality is that monitoring and tracking plastic is a problem that will have to be approached from numerous angles. To cover litter from large pieces to microplastics, from known to unknown sources, from the surface to the seabed, a strategic combination of technologies will offer the best information. Paired with a geopolitical analysis of the primary sources of plastic, this will allow obtaining an accurate model of plastic behavior.

Combining all these technologies and approaches will result in an extensive data bank. A challenge will arise in managing this data in terms of size and because it will come from several separate sources. For this reason, AI is also expected to play a vital role in the fight against plastic pollution (Chidepatil, et al, 2020). AI is the computer's capacity to perform tasks generally attributed to intelligent beings (Copeland, 2020). This technology's effectiveness in gathering useful information from large data sets makes it an ideal supporting tool for the sensing technologies described above. AI has already been used in the context of ocean monitoring. Scientists at the Charles Stark Draper Laboratory and the New England Aquarium have successfully used AI to create a probabilistic model of whales' locations in the oceans (Kako, Morita and Taneda, 2020). It is hoped that the technology will be used similarly to produce statistical predictions of plastic location and behavior using data gathered from both space-based and terrestrial sources.

The results of Biermann, et al. (2020) also show high effectiveness of machine learning. As stated previously, the accuracy of this approach reaches eighty-six per cent and can be further scaled. Despite this, the other side of AI is that powerful computing resources should be engaged. The best way to mitigate this limitation is to find a way to quickly analyze the images without the engagement of machine learning algorithms, using them on a small amount of data. By analyzing NDVI images from Sentinel-2, NDVI indirectly demonstrates the evidence of pollutants on the water's surface. This is because several plastic patches create favorable conditions for seaweed. Therefore, it should be additionally researched whether this fact can be leveraged to reduce possible places to be analyzed.

### 2.4.3 Non-space-based Mitigation Techniques

There are two approaches to non-space-based ocean plastic mitigation: technical and non-technical techniques. Notable terrestrial approaches include a variety of innovative technologies which illuminate the keen interest by many environmentalists to tackle ocean plastic pollution. To help keep track of such endeavors, Duke University generated a bevy of technologies in their 'Plastic Pollution Prevention and Collection Technology Inventory' which includes 52 technologies aimed at either reducing or mitigating ocean plastic pollution, delineated by their application on either macro or micro plastics, or both (Schmaltz, et al, 2020). An example of a promising technology focusing on macroplastics has been developed by The Ocean Cleanup Project (OCP), which has been posited to clean up 7,032 U.S. tons of debris annually (Slat, et al, 2014). Their system uses a series of barriers (i.e booms) and platforms (used in mooring). However, important to keep in mind is what Cordier and Uehara put forth regarding how 'much' technology of this kind would be needed in order to significantly reduce ocean plastic levels. Their assessment claims that it would take 1,924 technologies used by OCP in order to decrease ocean plastic to levels seen around 2010 (Cordier and Uehara, 2019). One can see the boom system setup shown below in Figure 15a and Figure 15b, provided by the Ocean Cleanup Project.

(a) *Boom system used in OCP (Ocean Cleanup Project, 2021).*

(b) *Close up of boom system used by OCP (Ocean Cleanup Project, 2021).*

**Figure 15:** *Display of booms used to collect plastic floating on the surface of the ocean.*

Another promising technological solution includes The Great Bubble Barrier (GBB), which works to siphon out of oceans and rivers both micro and macro plastics via a screen of bubbles created by pumping air through tubes on the river and ocean floors (The Great Bubble Barrier B.V, 2020). It is scalable, that is, versatile, and can be adapted to a variety of marine environments. A picture of the GBB is shown below in Figure 16.

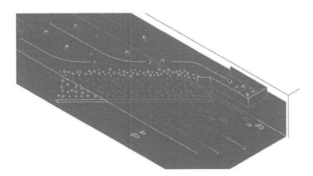

**Figure 16:** *How the GBB functions. Air is pushed through tubes on the ocean floor. This forms bubbles which rise and sift debris (including plastic) into bins for collection (The Great Bubble Barrier B.V, 2020).*

However, non-technical, yet terrestrial mitigation techniques are important too. These are tailored to inspiring more local and communal forms of action. Good Karma Projects and the Surfrider Foundation Europe are good examples of this. They aim to educate children on the importance of environmentalism via outdoor presentations, workshops and activities. On a larger scale, businesses and corporations have gotten involved in ocean plastic mitigation and cleanup, attempting to reduce their plastic footprint (Heidbreder, et al, 2019). To augment things further, municipalities have gotten involved too, such as the Italian municipality of Capannori, which became the first municipality in Italy to adopt Zero Waste goals (van Vliet, 2013).

Finally, key players of environmental endeavors have certainly made their claim on addressing the issue. These include organizations like NOAA and the European Space Agency (ESA). NOAA has adopted a Strategic Plan for 2021-2025, outlining how both they and their partners will aim to have an ocean free of marine debris; their goals include prevention, monitoring, detection, as well as research and removal (NOAA, 2020). ESA, on the other hand, has set up what is called, 'The ESA Initiative', which provides space start-ups with a platform to further develop their ideas for leveraging space technology for ocean plastic detection and monitoring (European Space Agency (ESA), 2019).

This overview has covered the context the Earth finds itself poignantly ensconced within in regard to ocean plastic pollution. The amount and location of ocean plastics was established in subsection 2.2. The consequences to biological entities of such pollution, and the social and economic context which informs it was discussed in subsection 2.3. Then, various approaches to ocean plastic detection, tracking and mitigation were discussed per the accentuation of space-based and non-space-based technologies, as well as the policy-based approaches. Now, attention shall be turned to this report's focus on an emerging geographic region in which ocean plastic pollution is a fast-growing issue, the Arctic.

## 2.5 What's Next? The Arctic Ocean

To date, there are five well-established gyres where ocean plastic accumulates. However, in an effort to contribute to the larger inquiry about the issue of ocean plastic, it is more useful to focus on a novel, less studied gyre, which has started to form in the Arctic. Too few have investigated ocean plastic and even fewer have focused on the ways in which space-based technology can be utilised to detect and monitor it. As such, this report will consider the Arctic as a case study.

The importance of a focus on this growing gyre of ocean plastic will be established first. This will include an explanation of the developing sixth gyre in the region. It will also consider relevant background information on the current geopolitical context in which the Arctic is ensconced (e.g. regional delineation of the Arctic by nation). Alterations to the Arctic Council will be presented as a way to effect collaboration across national borders. This section will also include relevant economic information about the impacts of plastic pollution on trade routes through the region, both contemporarily and in the future.

Next, an application of space-technologies to detect and monitor plastic in the Arctic waters will be ascribed. Proposing space-based approaches to detect and monitor plastic within this region will allow stakeholders to start collecting data on where it is travelling and accumulating. Doing so will provide the necessary proof required to hail mitigative action by governments and organizations. Such intervention would prevent the emerging Arctic plastic gyre from turning into the GPGP in terms of magnitude. This report's contribution to this vision sets the foundation for this to occur. It will accentuate the potential need for augmentation of non-Arctic ocean plastic space detection technology. Special attention will be given to how such technologies can be adapted for conditions of extreme areas such as the Arctic, which may prove useful in other extreme environments too.

## 2.6 Chapter Summary

The majority of marine plastic will remain trapped in the gyres for decades to come, and given the annual rise in plastic production and environmental losses, the negative environmental, social and economic impacts of plastic are important. An additional 28-71 million tons of plastic are estimated to have been added into the marine environment from land-based sources between 2011 and 207 (Beaumont, et al, 2019). Plastic pollution can be tracked and monitored using non-space based techniques, such as modeling approaches to river monitoring. Plastic can also be monitored using space and airborne technology such as NIR spectroscopy. Active GPS trackers map plastic routes and Sentinel-2 satellites are used by many to track and monitor it. Current solutions allow for the identification of plastic patches in oceans and rivers with an accuracy of eighty-six per cent. Marine debris has started as an anthropogenic epidemic, thus human-driven solutions need to be part of the solution. The key to this is an interdisciplinary and collective international and intergovernmental approach which could pave the way for a brighter future.

The next chapter discusses more in depth the effects of plastic pollution in the Arctic Ocean. Specifically, the chapter will focus on Arctic sea ice and its interactions with local climate, as well as the broader effects of changes in the Arctic sea ice cover and its release of debris.

# 3
# The Tip of the Iceberg:
# Plastic Pollution in the Arctic Ocean

'For most of history, man has had to fight nature to survive; in this century he is beginning to realise that, in order to survive, he must protect it.' - Jacques-Yves Cousteau, (French naval officer, ecologist, scientist, researcher).

## 3.1 Introduction

Plastic pollution has been documented in all marine environments, from coastlines, ocean surfaces (Barnes, et al, 2009), within deep-sea sediments, on the ocean floor and even lodged in sea ice (Obbard, et al, 2014). The most well-documented measurement of plastic pollution in oceans has focused on buoyant plastic floating at the surface leaving the full extent of debris accumulation not fully determined. Major figures in plastic pollution studies Eriksen, et al. (2014), Lebreton, Greer and Borrero (2012) and Cózar, et al. (2017) have estimated floating plastic in ocean gyres using circulation models to stress the importance of better understanding the transformation process of plastic in seawater. Sebille, England and Froyland (2012), goes as far as to say that a sixth gyre is forming in the Barents Sea through inter-ocean exchanges enabling patches to 'leak' into one another.

This chapter aims to provide an introduction to the Arctic Ocean and the Arctic nations impacted by plastic pollution. It first describes the importance of preserving the Arctic Ocean from plastic pollution, and in general, the effects of climate change on the region. Furthermore, the section provides an investigation into the effects of ocean plastic pollution on indigenous populations, new opportunities for Arctic shipping routes, and potential increases in plastic pollution from greater marine activities. The investigation on Arctic shipping routes will determine whether or not a correlation exists between increased Arctic activities via new shipping routes and increases in plastic pollution generated in the area. Lastly, it considers if an Arctic treaty is necessary to coordinate environmental preservation efforts. Professor Mariana Mazzucato, Director of University College London (UCL) Institute for Innovation and Public Purpose (IIPP), stated that, 'far too many governments have become passive lenders of last resort, addressing problems only after they arise...it costs far more to bail out national economies during a crisis than it does to maintain a proactive approach to public investment' (Mazzucato, 2021). There is a great opportunity here for the public sector to champion plastic pollution efforts in the Arctic.

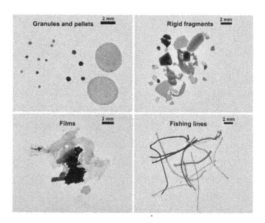

**Figure 17:** *Microplastics found in the Arctic Ocean. Since a diverse sampling of plastics was found, researchers do not believe that the plastics originate from the local populations (Cózar, et al, 2015).*

Plastic pollution was first reported in the Arctic in the 1970s following observations of plastics in the Bering Sea (Merrell, 1980). Following a two-year period survey (1972-1974) during which plastic on beaches around the Amchitka Island were collected, plastics had increased in weight by nearly sixty per cent per year, which translates to an increase from 2,221 pieces to 5,367 pieces (Halsband and Herzke, 2019). The majority of the plastic documented had come from local fishing vessels but most significant was that some pieces had traveled over 1,100 km from the Asian coast (*loc. cit.*). This was one of the first realizations that plastic pollution spreads on a global scale. Examples of microplastics found in the Arctic Ocean are in Figure 17. The Arctic is highly connected to other bodies of water through the Fram Strait, Bering Strait, Alaskan Archipelago, North Sea, and North Atlantic current. This accentuates the many marine avenues through which plastic end up in the Arctic Ocean.

The many avenues for plastic to reach the Arctic, particularly per Earth's thermohaline circulation, establish it as a 'dead end for plastics' (Cózar, et al, 2017). Wind-driven ocean circulation patterns are what generate these so-called 'gyres' that expand when more debris enters circulation. As mentioned previously, one of the many reasons this report focuses on the Arctic Ocean is because plastic pollution is a developing crisis in this area, and with proper monitoring and tracking, governmental coordination and investment, it could be avoided before it becomes as significant as one of the other gyres.

### 3.2 Importance of the Arctic Territory

The Arctic Circle, one of the five major circles of latitude, encompasses eight Arctic states which make up the territorial land and sea claims, international waters (not owned by any state), and exclusive economic zones of the area (Arctic Council Secretariat, 2021; Craig, 2016). These countries are Canada, the United States (per Alaska), Norway, Russia, Sweden, Denmark (per Greenland), Finland and Iceland (Figure 18).

**Figure 18:** *Arctic Ocean States (Adobe, 2021).*

As of late, the Arctic Circle has gained attention from climate activists, academics, indigenous and local populations, global climate forums and world leaders. This is due to global temperatures rising, graver predictions of rapidly melting Arctic sea ice and now plastics pollution beginning to slowly appear in the region from the Earth's thermohaline circulation moving plastics 'upwards.' The Arctic Circle has an abundance of natural resources, seaports, growing tourism, marine wildlife, and research opportunities, all of which have proven to be vital for these countries' livelihoods; therefore, the increasing amount of plastic pollution in this region puts these opportunities at risk. One of the risks is an ice-free summer, which is not a misnomer. Pristine habitats with rare wildlife and native communities have existed for thousands of years, but with an ice-free summer, climatologists fear that fossil fuel companies will attempt to avoid pursuing alternative energy sources and drill the Arctic (Council on Foreign Relations, 2021).

Scientists predict an ice-free summer to come as soon as 2035 (British Antarctic Survey, 2020), thus making it pertinent that mitigation techniques are put in place now to preserve the Arctic. With economic expansion both regionally and internationally, greater human activity and global climate change threaten marine ecosystems and the populations that depend on them.

### 3.2.1 The Heterogeneity of Arctic Populations

Approximately four million people live within the Arctic region throughout the year, all of whom depend on marine ecosystems; 12.5% of them are indigenous to over forty ethnic groups (Bundesministerium für Bildung und Forschung, 2018). Figure 19 depicts the total population, Arctic population, Arctic land area, and the number of ports each state has. Harsh climate and less frequent accessibility to major urban centers by people living in the region fosters an increased reliance upon marine ecosystems for their livelihood, especially regarding food supply.

## ARCTIC STATES

**Alaska**
Population: 731,545
Arctic Population: 731,545
Arctic Area: 0.55M km$^2$
Arctic Ports: 9

**Norway**
Total Population: 5.33M
Arctic Population: 393,000
Arctic Area: 96,225 km$^2$
Arctic Ports: 4

**Russia**
Population: 144.4M
Arctic Population: 2M
Arctic Area: 4.28M km$^2$
Arctic Ports: 12

**Canada**
Population: 37.59M
Arctic Population: 200,000
Arctic Area: 3.99M km$^2$
Arctic Ports: 2

**Finland**
Population: 5.52M
Arctic Population: 180,000
Arctic Area: 111,685.2 km$^2$
Arctic Ports: 1

**Iceland**
Population: 356,991
Arctic Population: 60-100
Arctic Area: Grimsey Island*
Arctic Ports: 1

**Greenland**
Population: 56,225
Arctic Population: 56,225
Arctic Area: 1.43M km$^2$
Arctic Ports: 1

**Sweden**
Population: 10.23M
Arctic Population: 552,420
Arctic Area: 67,544.25 km$^2$
Arctic Ports: 4 port hub

**Figure 19:** *Arctic Demographics (2019). Russia, Greenland, and Canada have the largest land area located within the Arctic Circle while Russia and Alaska have the most ports. (information gathered from ports.com and statista.com. Grimsey Island straddles the Arctic Circle total land area 5 km2 undetermined area inside Circle).*

This area is of great global interest because of its abundance of natural resources (e.g. oil, gas, hydrocarbons), fisheries, seaports, growing tourism, marine life and research opportunities. Arctic Circle activities are slowly expanding as sea ice melts, opening up greater vessel-accessibility, new underwater areas to explore and new land activities which could bring more people to the region (Halsband and Herzke, 2019). However, this will consequently generate more plastic pollution in the area, too. The Scandinavian countries and Alaska primarily use the Arctic Ocean for fisheries, shipping, oil and gas extraction, marine tourism, and trade routes. Alaska also uses it for launching research activities. Iceland is highly active in fishing, marine tourism and trade, while Greenland uses the ocean for hunting, fishing and trade. Canada is involved in hard minerals extraction, research, oil and gas extraction and trade. Lastly, Russia, through its largest Arctic port, Murmansk, uses the Arctic for trade, and extraction of oil, gas and hard minerals. In section 3.3.2, is a greater discussion regarding the importance of marine ecosystems for indigenous populations that depend on a pristine Arctic.

The Arctic region climate is also important for polar species such as seals, walruses, bears, ice algae and different crustaceans relying on the unique climate and existence of sea ice to survive (Folger, 2017). When ice melts it takes away surfaces that walruses and polar bears use for rest, mating, and when migrating to different areas for food. In 1980 there were 7.01 million square kilometers of sea ice but as of 2015, only 3.88 million square kilometers refroze, resulting in polar species habitat loss (Starr, 2016). Other than impacts on human populations and biodiversity, there are several other reasons why preserving the Arctic Ocean and Arctic sea ice from plastic pollution is important. These include ice-albedo feedback and thermohaline circulation.

### 3.2.2 Ice-Albedo Feedback Loop

Albedo is a metric to calculate the reflectivity of light surfaces. Light or white surfaces do not absorb or retain as much heat as darker surfaces, thus leaving them cooler (Misachi, 2017). Open water reflects roughly six per cent of solar radiation while sea ice reflects fifty to seventy per cent in the summer months. With the addition of snow on top of sea ice, the albedo significantly rises reflecting up to ninety per cent of incoming solar radiation (National Snow and Ice Data Center, 2020c). This feedback is important in regulating Earth's temperatures. As sea ice melts, it reduces the number of surfaces for light to reflect off of, which exacerbates further sea ice melting; this subsequently leads to ocean warming known as the ice-albedo feedback.

During sea ice growth, debris and plastic pollution get trapped inside the ice. This accumulation further diffuses reflections of sunlight, increasing the albedo and promoting further surface melt (Galley, et al, 2015). Although the presence of plastic does not hinder the growth of the ice, it does interfere with the natural processes and salinity. Geilfus, et al. (2019), found that high concentrations of microplastic in the Arctic caused changes in sea ice albedo. Furthermore, surface plastic could affect light penetration depths impacting photochemical, photobiological and biogeochemical processes in the ice and seawater.

### 3.2.3 Thermohaline Circulation System

Earth's thermohaline circulation system is a natural heat-and-saline-driven process that originates from the polar regions as ocean water gets very cold and freezes over, thus causing surrounding seawater to get saltier, increase in density, and sink (National Oceanic and Atmospheric Administration, 2020c). The system uses ocean currents for transporting heat from the equator to the polar regions. Warm water is transported away from the equator northward while cold water is transported southward. As the warm water goes northward it becomes more dense and sinks and as the water heats up going southward it rises near the equator creating a temperature balance. The United Nations Environmental Program (UNEP) stated, 'when the strength of the haline forcing increases due to excess precipitation, runoff, or ice melt, the conveyor belt will weaken or even shut down' (GRID Arendal, 2007); therefore, the melting of the Greenland ice sheet provokes jet stream changes as more cold water enters the system. The thermohaline circulation system also acts as a conveyor belt for plastic pollution to travel upwards and get caught in the Beaufort Gyre (Cózar, et al, 2017). The following section will discuss the formation of the sixth gyre explaining how Earth's natural processes are creating a dead-end for plastic pollution in the Arctic Ocean.

### 3.2.4 A Sixth Gyre is Forming

The Arctic Ocean and the polar ice cap are governed by the natural processes of the Coriolis Effect-induced wind and water circulation, thermohaline currents and the bordering landmasses. These forces result in the Beaufort Gyre and the Transpolar Drift (Figure 20) that cycle sea ice and debris across the Arctic.

Towards the Eurasian side of the Arctic Ocean, the movement of ice and water is driven by the Transpolar Drift. This movement takes surface ice very near the North Pole and was used by explorers such as Fridtjof Nansen for 'over ice' transport across the ice cap. Both Nansen's *Fram* expedition (for which the body of water between Greenland and Svalbard is named) and the wreck of the USS Jeannette were transported from the East Siberian Sea to Greenland over the course of roughly three years (Nansen, 1897).

Figure 20: *Graphic from the Protection of the Arctic Marine Environment (PAME) showing the patters of ocean currents and surface winds that drive the Beaufort Gyre and the Transpolar Drift. Adapted from (Protection of the Arctic Marine Environment, 2020).*

The Beaufort Gyre cycles ice and water closer to the American continent side of the Arctic Ocean. Historically, this is where the oldest ice in the Arctic could be found (Lei and Wei, 2020). However, changes in the air and water currents have caused the loss of up to ninety-five per cent of the 'old ice' over four years in age between 1985 and 2018 (*loc. cit.*). Years of strong ice growth can reverse years of losses but the 'new ice' is not a sufficient replacement. The newer ice tends to contain higher concentrations of salt (and potentially anthropogenic debris) that allow the ice to melt more easily each spring (Comiso, 2012).

Air and wind interactions with surface currents appear to be a main driver for the migration of sea ice from the Arctic to the Greenland Sea (Ma, Steele and C. M. Lee, 2017) and are accelerating the circulation of the Beaufort Gyre more when the ice retreats (Armitage, et al, 2020). These wind-water interactions are stymied by heavy ice coverage but melting ice allows for greater impacts of the Arctic air currents on the Beaufort Gyre's motion, including the Transpolar Drift current (Ma, Steele and C. M. Lee, 2017). As these patterns gain more energy, they allow for accelerated ice loss and stronger currents which draw debris from the North Atlantic and North Pacific Oceans, two of the most polluted in the world (Eriksen, et al, 2014).

The same wind-driven surface currents also appear to be responsible for the large volume of lightweight foam plastics that are driven out of the GPGP, causing it to have debris of higher average density than any of the world's other oceans (Lebreton, et al, 2018). The wind and shallow-water currents are believed to be to blame for the small and lightweight plastics 'missing' from the GPGP appearing along the Alaskan coast (*loc. cit.*). In these ways, an understanding of wind-driven currents is integral to characterizing plastic pollution in the Arctic Ocean.

Arctic plastic is found to be smaller in size and in the form of foam-type plastics more often than in other oceans (Cózar, et al, 2017). The coastal population density of the Arctic is low compared to the North Atlantic and Pacific, which suggests that pollution is being fed into this region from the more polluted oceans as warmer surface waters are pushed north (Cózar, et al, 2017; Polasek, et al, 2017). Other explanations for the small particle sizes can be traced to microplastic particles found in the Arctic snow, likely from thousands of kilometers to the south, which are carried on north-moving air currents (Bergmann, et al, 2016). In turn, the plastics which accumulate on the ice caps bring the risk of increasing their albedo and further increasing ice melt (Bergmann, et al, 2016; Cózar, et al, 2017).

The Fram Strait in particular allows for the colder, fresher water to flow out of the Arctic Ocean and warmer surface water to rush in bearing debris. The Fram Strait is a unique opening to the Arctic due to its sheer size and depth (Klenke and Schenke, 2002). The increase in melting ice leads to an increase in the power of nearby currents, thus driving away more ice from the polar cap and consequently allowing for more melting. The Fram Strait also appears to be the main route by which sea ice leaves the Arctic Ocean (Comiso, 2012).

Water cools as it travels north from the Pacific and Atlantic, becoming denser as it approaches the North Pole. This causes more debris to float to the surface towards the Arctic (Thevenon, Carroll and Sousa, 2014), resulting in more debris that becomes trapped in polar ice. Anthropogenic debris can remain suspended in the water column for an extended period of time; these plastics are often discounted by researchers, but the mechanics of the Arctic Ocean's dynamic cooling and warming of water currents may better allow such plastics to be accounted for since their movement is better characterised per thermohaline circulation assessment.

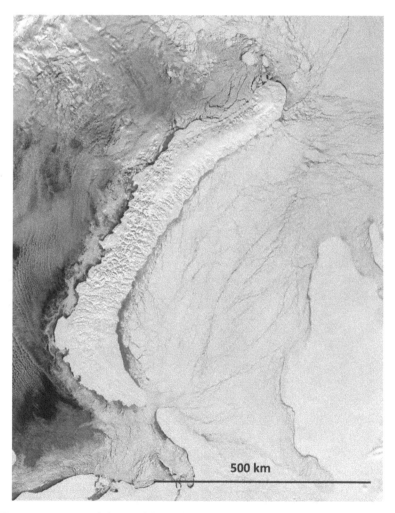

**Figure 21:** *Image from VisibleEarth.NASA.gov shows Novaya Zemalya, the Northernmost stretch of the Ural Mountains extending into the Arctic Ocean. Here, sea ice can be seen crashing against this prominent landmass on a grand scale, slowing the movement of the Transpolar Drift (National Aeronautics and Space Administration (NASA), 2004).*

Arctic ice is believed to hinder the progress of debris flowing into the Arctic Ocean, though meltwater can wash debris into the ocean and glacial tilling can also 'scrape' it from the land (Kylin, 2020). Reductions in ice for longer periods of time throughout the year will lead to more water circulation that draws more debris from the North Pacific and Atlantic (Polasek, et al, 2017). This appears to be the main mechanism by which debris accumulates on remote Arctic beaches and 'ice floes' (*loc. cit.*); the term 'ice floes' is jargon for large pieces of floating ice of at least 20 meters across at the widest point (National Snow and Ice Data Center, 2020c).

The Ural Mountains extend into the Arctic Ocean as the Novaya Zemlya archipelago (Figure 21) and the Russian Federation's northernmost inhabited territory. The mountains help slow the anticyclonic motion of water and ice in the Arctic Ocean (Cho and Kim, 2021). As Arctic Ocean wind and wave action are anticipated to increase later this century (Ryan, et al, 2020), Novaya Zemlya will be subjected to increased erosional forces. This may reduce its effectiveness in stymieing the growing gyre. The retreat of ice in Arctic spring and summer will also allow for more flow between the ice cap and the archipelago, likely increasing the circulation of warmer waters into the Arctic Ocean and further reducing the lifespan of formed ice (Comiso, 2012).

Arctic shipping, fishing and tourism are already large potential drivers of pollution: the Western Arctic's high density of fishing buoys can easily be traced back to the United States' largest fishery, which is located in the Bering Sea (Polasek, et al, 2017). Discarded fishing equipment is the second most common type of debris found in the Arctic Ocean (Kylin, 2020). The formation, migration and melting of Arctic ice also appears to be the cause of anecdotal incidents where terrestrial mammals are found dead and tangled in fishing gear surprisingly far inland (*loc. cit.*). To review, what has been shared, the Arctic is already a 'dead end' for marine debris (Cózar, et al, 2017); Melting ice results in these driving currents growing stronger, fueling additional ice losses and debris accumulation which establish the formation of the sixth garbage gyre.

### 3.3 Socio-economic Framework In The Arctic

#### 3.3.1 Economic Opportunities in an Ice Free Arctic

As stated previously, the Arctic Ocean is one of the most rapidly evolving regions when considering ocean plastic pollution; Arctic ice thinning and shrinking has increased opportunities for maritime activities, thus enabling access to previously inaccessible areas (Silber and Adams, 2019). Arctic maritime traffic management and navigation have faced tremendous challenges due to the physical properties and natural motions of sea ice (Council on Foreign Relations, 2021). Fluctuations in ice ultimately determine the time window in which marine activities take place. Despite diminishing sea ice levels, there are still threats from the subzero temperatures that hinder human capabilities (e.g. harsh working conditions) and machine-functionality (e.g. icebreakers and onboard mechanisms), especially during the winter months.

Today, the warming Arctic waters caused by climate change are literally opening up opportunities for shipping industries to use the northern shipping routes but researchers warn that increased activities may further exacerbate already existing environmental problems, including plastic pollution if left unmonitored and unregulated (Cózar, et al, 2017; Katz, 2019). Shown in Figure 22 is the shrinking of the sea ice delineating the Northern Sea Route (NSR) and the Northwest Passage (NWP). Most significant is the creation of the North Pole Route, also known as the Transpolar Sea Route (TSR) or North Pole Route. This route could enable greater marine activities, faster routes, fuel savings and greenhouse gas reductions. Having said this, the choice to choose the Arctic was specifically made because it is an emerging locale of plastic pollution deposition.

At the 2011 Arctic Forum, then Prime Minister Putin stated, 'The shortest route between Europe's largest markets and the Asia-Pacific region lies across the Arctic. I want to stress the importance of the Northern Sea Route as an international transport artery that will rival traditional trade lanes in service fees, security, and quality' (Gosnell, 2018). This exemplifies growing national interest in expanding the use of northern routes. Over the last six years there has been a twenty-five per cent increase in the amount of ships navigating the Arctic from 1,298 ships in 2013 to 1,628 in 2019 (Protection of the Arctic Marine Environment, 2020). These were mostly fishing vessels.

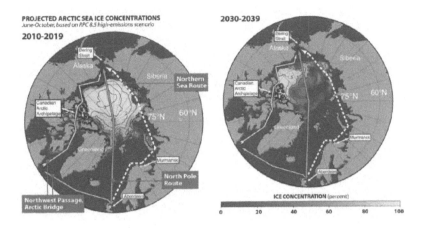

**Figure 22:** *New Arctic Trade Routes - As see ice melts, new trade route opportunities are opened potentially increasing plastic pollution. The most significant route is the Transpolar or North Pole Route that could become heavily traversed to cut travel time and fuel spending (Aksenov, et al, 2017).*

The traditional East Asia to Northern Europe shipping route through the Suez Canal going from Yakoma to Rotterdam takes 11,200 nautical miles (20,742 km) but only 6,500 nautical miles (12,038 km) through the Transpolar Sea Route shortening travel times by twelve to fifteen days, a significant economic benefit (Kuperman, 2014). In Table 3 are calculations on distances saved and time saved traveling through the TPR instead of the Suez Canal. Both the China Ocean Shipping Company (COSCO) and United States Senator Angus King of Alaska expressed the significance of this route as a way to connect the Atlantic and Pacific Oceans through direct linear passage over the Arctic Ocean (*loc. cit.*). Reducing the risk for error between long distances could improve security standards and centralizing polar trans-shipment operations could further cut travel times. Variables such as account fees, freight rates, operational costs and using larger bulk carrier and shipping vessels can further streamline efficiency and reduce costs. Murmansk, Kirknes and East Icelandic ports have unique opportunities to be major shipping hubs to service the Northern Sea Route due to their major geographic and economic advantages.

**Table 3:** *Sailing Distances by taking the TSR (Humpert and Raspotnik, 2012).*

| Port of origin | Port of destination | Distance in nautical miles | | Days at sea at 17 knots | | Distance savings in % |
|---|---|---|---|---|---|---|
| | | Via Suez Canal | Via TSR | Via Suez Canal | Via TSR | |
| Tokyo | Rotterdam | 11,192 | 6,600 | 27.4 | 16.1 | -41 |
| Shanghai | Rotterdam | 10,525 | 7,200 | 25.8 | 17.6 | -32 |
| Hong-Kong | Rotterdam | 9,748 | 8,000 | 23.9 | 19.6 | -18 |
| Singapore | Rotterdam | 8,288 | 9,300 | 20.3 | 22.7 | +12 |

Melting sea ice has already prompted NSR sailing times to decrease from 20 days in the 1990s to 11 days in the 2010s (Aksenov, et al, 2017). The economic potential of these routes paired with the adoption of super-slow sailing has already saved shipping companies more than £68M ($90M) on just fuel savings (Vidal, 2010), thus elucidating a significant cost saving opportunity (Humpert and Raspotnik, 2012). Instead of navigating faster with the absence of ice, many vessels have been fitted to operate at super-slow speeds resulting in a reduction of emissions. With the TSR time savings, ships could adopt 'slow steaming' (12 knots) to save fuel, but still arrive at the same time as if travelling through the Suez Canal, thus reducing emissions (Sch0yen and Bråthen, 2011). With reduced shipping speeds, greenhouse gas emissions are also reduced providing further environmental savings.

Despite these benefits, an increase in plastic pollution may occur as large contributors to plastic pollution are from fisheries and land based activities that would increase from these new trade routes. Historically, fishing has contributed to immense amounts of plastic pollution. On average, global discharge of fishing gear into the oceans exceeds 640,000 tons, which comprises approximately ten per cent of global marine plastic pollution (Environmental Investigation Agency, 2020). Moreover, vessels represent a moving source of plastic that is not easily traceable unless the vessel is monitored full time. In the Arctic, plastic pollution poses a unique problem because of how dependent northern populations are on the marine ecosystem for their food and culture. Peter Murphy, Alaska's regional coordinator of NOAA's Marine Debris Program stressed that plastic pollution has, 'a much more straight-line impact,' in the Arctic region (Katz, 2019).

Studies show that remote Alaskan coastal hotspots have similar plastic pollution levels as highly populated urban regions (Whitmire and van Bloem, 2017).

There is evidence that the economic centers of European and Asian countries are slightly moving northwards (Verny and Grigentin, 2009), to increase accessibility to trans-Arctic shipping routes. Roughly eighty per cent of all plastic pollution comes from land-based surfaces, signifying the figures in the Arctic could increase, shifting this estimate. Being able to track and monitor plastic pollution in the Arctic before the gyre grows at rates similar to more developed gyres could bring significant savings in needing to spend money on ocean cleanup projects.

### 3.3.2 Correlation Between Plastics and Shipping Routes

The connection between shipping and plastic debris is presently poorly understood but remains an undeniable contributor to the global crisis. Commonly cited statistics claim that eighty per cent \ of plastic debris originates from land-based sources. However, there is a plethora of evidence that contradicts this, since studies have observed that fishing gear such as nets and lines are the most commonly identified macro debris (Moy, et al, 2018), and compose approximately forty-six per cent of the GPGP mass (Lebreton, et al, 2018); fishing buoys are the greatest contributor to plastic debris mass (Eriksen, et al, 2014). Therefore, sources of plastic pollution from ships should be given proper attention.

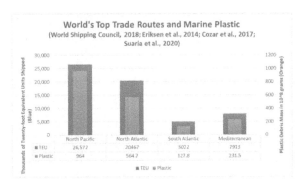

**Figure 23:** *World's Busiest Trade Routes. Available data from multiple sources was combined to show a disturbing trend between the world's shipping routes and plastic pollution in the oceans. This trend should also be seen as a reflection of coastal populations and fishing practices, not simply shipping practices.*

Shown in Figure 23, the World Shipping Council (2017) estimated the number of twenty-foot equivalent units (TEU), a common measure of marine cargo, being shipped across the world's busiest trade routes. Combining the TEU of the multiple trade routes that cross each ocean reveals a strong correlation between the world's busiest trade routes and the mass of plastic in those oceans. This correlation should not be seen as strictly causative since marine trade is also largely driven by population density and industrial development in coastal areas, meaning that marine traffic also correlates with other leading causes of marine pollution.

Equipment such as gloves and fishing storage boxes are also contributing to the mass of debris considerably in Scandinavian waters (KIMO International, 2019a). Scandinavian nations and the populations of the Arctic are very dependent on sustainable fishing practices due to the environment which limits agricultural food sources. However, illegal and environmentally harmful fishing methods are the cheapest practices in the short-term. Fishing gear that is exposed to the marine environment for too long cannot be effectively recycled after recovery (*loc. cit.*). Furthermore, plastic marine debris and illegal fishing practices not only harm the environment but the people living in coastal communities too (Lüber and McLellan, 2017). Paint used on ocean-going vessels is coming to light as a potential source of micro and nano debris in the oceans (Mannaart, et al, 2019). Harsh sunlight, saltwater and high winds can cause these paints to 'flake' and fall off the ship, resulting in debris.

Shipping containers are commonly lost in small numbers and occasionally in catastrophic mass loss events. 'Containers lost at sea create both an environmental and a shipping hazard' (PortandTerminal.com, 2020). The Ilulissat Declaration of 2008 also acknowledges the risk that shipping accidents can cause 'irreversible disturbance' to sensitive ecologies (Dodds, 2013). The World Shipping Council (2020) estimates that an average of 1,382 containers are lost at sea out of the hundreds of millions of containers shipped each year. This small number of lost containers presents an environmental hazard due to the contents of their cargo. The World Shipping Council is advocating for revisions to existing laws to create global standards for container construction, weighing and packing. The lack of such standards enhances opportunities for cargo to be lost. 'Catastrophic losses' are defined as single incidents that result in fifty containers or more being lost and the comparatively low number of catastrophic events represents the majority of losses (World Shipping Council, 2020).

Improper loading of cargo presents severe danger to people, vehicles and goods. In Bandar Abbas, Iran, a reported 153 containers were 'suddenly submerged' on March 18th, 2019 from the deck of a small cargo ship that capsised while being loaded (Vessel Finder, 2019). Improper operation of certain vessels also causes lost containers. The M/V Arvin, a light cargo ship built by the Soviet Union in 1975 'broke in half' (The Maritime Executive, 2021) on January seventeenth, 2021 in the Black Sea. The vessel was reportedly designed for river use and had numerous safety violations in 2020 (Voytenko, 2021). An unknown number of crew died or still remain missing. The incident is reminiscent, though smaller in scale, of the sinking of the MOL Comfort in June, 2013, when the vessel broke in half and lost more than 4,000 containers in the Indian Ocean (World Shipping Council, 2017).

More modern ships like the MSC Zoe, self-proclaimed as being the largest container vessel operating in Europe, are designed to 'roll' with the waves to reduce the chances of losing the ship. However, the torsional forces generated by rolling aboard these massive ships is sometimes too much for the stacks of containers onboard, which can collapse or fall overboard, as seen in the case of the MSC Zoe losing 342 containers on January 1st 2019 (Maritime Research Institute Netherlands, 2020).

Shipping containers are claimed to have been responsible for the damage of several recreational vessels, while others insist on a natural explanation such as collisions with whales (International Whaling Commision, 2019). 'But it could be just a couple of inches above the surface, and if it's filled with the right amount of polystyrene, it could stay there forever,' Toby priestly says regarding lost shipping containers (Fretter, 2017). Polystyrene and other plastic used for packaging purposes are a large portion of the single-use plastic produced (an estimated 287 metric tons of plastic produced globally went to packaging out of 380 metric tons estimated made in 2015) (Geyer, Jambeck and Law, 2017). Sailors are often unable to identify what their vessels collide with, leaving a great deal of uncertainty on what the causes and effective mitigation of these incidents could be.

Plastic debris is a hazard to ships and fisheries (Thevenon, Carroll and Sousa, 2014). Debris can cause damage to equipment and vessels, injure crews and occupy their time with cleaning and repairs. Smaller recreational, shipping and fishing vessels are at severe risk of catastrophic damages from anthropogenic debris (Fretter, 2017). Ghostnets can cause engine failures; more solid objects can destroy keels and rudders and create hull breaches.

In 2008, the United Kingdom recorded 286 sea rescues due to ship propulsion systems being tangled in debris (Niaounakis, 2017). The same also reported $4.3 million USD in lost fishing revenue due to plastic debris. Plastic damages not only hulls, rudders, keels, sensors, water intakes (especially on fishing vessels that refrigerate cargo) but also damages their plastic nets (Carr, 2019).

Executive Secretary for KIMO, Mike Mannaart, estimates that the Fishing for Plastic Fleet can clean roughly 300 tons of ocean litter at a cost of approximately 100,000 EUR (~119,000 USD) per year (KIMO International, 2019b). MSC Zoe's single container loss event resulted in 9,000 tons of litter on January 1st, 2019. The sudden input of debris prompted a response not just from quickly overwhelmed local volunteers but also from local coast guards and armies (BBC, 2019), transferring the cost of pollution directly to the countries affected. The United States National Ocean Service (NOS) and NOAA have set aside $5 million USD for 'Marine Debris Prevention and Removal grants in FY2021' with the goal of creating proactive solutions to reducing marine debris, particularly from fishing equipment (NOAA, 2020).

In 2018 the first commercial ship was able to cross the Arctic unassisted by icebreakers (Nordregio, 2019). Should transport routes through the Arctic increase, pollution, both intentional and accidental, will increase greatly. Private vessels alone are believed to contribute to twenty-five per cent of the marine-sourced debris (Stachowitsch, 2019).

### 3.3.3 The Influence of Plastics On Indigenous People

The indigenous people in the Arctic circle have a distinctive status given by the Arctic Council. In fact, the Arctic has a particular political framework since it is a 'zone of peace and cooperation' (Fondahl, Filippova and Mack, 2015). Indigenous people, although living in isolation from Western society, are indirectly affected by it.

Indeed the most noticeable change has been the changes in the health of Inuits, especially in the last fifty years of the twentieth century. Whilst the proportion of infectious diseases has greatly reduced, it is still high compared to the Western average, and the Inuit community suffers from many chronic diseases which can be linked back to social, lifestyle, and environmental health (Bjerregaard, et al, 2004).

The Indigenous Peoples of North-Western Siberia have a diet that comprises traditional foods, typically fish or reindeer. However, due to climate change and plastic pollution, the fishing season and the migration routes have been thrown into disorder, which leads to a decrease in fish consumption.

Andronov, et al. (2020), did a screening of 985 indigenous people from Yamal-Nenets Autonomous Okrug for five years and looked at the quantity in acquisition and consumption of fish that they had. They concluded that the duration of the seasons in which they eat fish is decreasing, therefore the quantity of fish consumed is reduced, too (*loc. cit.*), which can lead to greater risk of negative health effects such as hypertension.

The Arctic and its inhabitants are adversely affected by plastics from regions of lower altitudes. These pollutants are shipped via the atmosphere, oceans and rivers that feed into the Arctic ecosystems and biomagnify through food webs. The microplastics are minute and can easily affect wildlife and fish species native to the Arctic. Thus, indigenous communities that rely on them as part of a traditional diet remain vulnerable to the potentially detrimental effects associated with the consumption of plastics (Monitoring and Program, 2018).

According to the 2018 Arctic Monitoring and Assessment Program (AMAP) reports, Arctic wildlife and fish experience various biological effects when exposed to potentially harmful levels of chemicals in the plastic. These include fluctuation in hormone levels, immune function, tissue and bone density, neurological and behavioral effects, DNA damage and reproduction harm, all of which can further increase complications in the human body as well (Monitoring and Program, 2018). Therefore, this can cause latent human health outcomes like damage to the developing brain, endocrine and immune systems (United Nations, 2009). Life forms at varying levels of the food chain are clearly impacted by chemical constituents of plastic.

The Arctic has over forty distinct tribes living within its territory. They include Saami in circumpolar areas of the following areas: Finland, Sweden, Norway, Northwest Russia, Nenets, Khanty, Evenk and Chukchi in Russia, Aleut, Yupik and Inuit (Iñupiat) in Alaska, Inuit (Inuvialuit) in Canada and Inuit (Kalaallit) in Greenland (University of Lapland, 2021). Indigenous people have developed a special connection to the land they inhabit and follow a traditional livelihood, such as reindeer herding, fishing and hunting.

The ocean has always sustained the livelihoods of many native communities around the world from the Arctic to the South Pacific. A study shows that coastal indigenous people eat nearly four times more seafood per capita than the global average and about fifteen times more per capita than non-indigenous people (Cisneros-Montemayor, et al, 2016). The plastic marine debris stress in the Arctic region poses a serious threat to the indigenous people's personal health and relationships with the sea that diversifies them as a distinct culture (Ota, 2017).

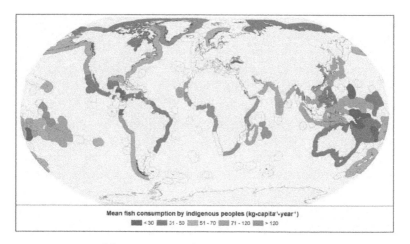

Mean fish consumption by indigenous peoples (kg·capita⁻¹·year⁻¹)
■ < 30  ■ 31 - 50  ■ 51 - 70  ■ 71 - 120  ■ > 120

**Figure 24:** *Mean fish consumption by indigenous people; 74 kg per capita (Cisneros-Montemayor, et al, 2016).*
*Greenland and Alaska have some of the highest mean fish consumption levels.*

Indigenous populations (Figure 25) such as the Kalaallit in Greenland, the Inuit and Inuvialuit in Canada, the Inupiat and Yupik in Alaska and the Yuit in Siberia are encountering adverse health outcomes. Studies indicate high levels of pollutants in Inuit mothers compared to the rest of the general population (Singh, Bjerregaard and Man Chan, 2014). In the Russian north coast, contaminant levels were measured in blood, cord blood and breast milk. Reports later indicated that mothers from the Chukotsy district had the highest contaminant-concentration in blood and breast milk; this was consistent in coastal Greenland and Northern Canada (*loc. cit.*).

**Figure 25:** *Extreme fishing diets of indigenous people in Greenland and Alaska (Hussin, 2019).*

Various studies organised by health category were taken up in the Arctic coasts regarding plastic pollution's effect on human health. The pediatric studies examined motherchild pairs and diagnosed the following positive associations with plastic and other pollutants: acute respiratory infections, immune status, neurological and behavioral functions, blood pressure, heart rate variability, thyroid functioning and attention deficit hyperactivity disorder (ADHD) (*loc. cit.*). This was followed by reproductive health studies that indicated pollutant associations with semen quality and reproduction hormones, sperm Y:X ratio and DNA damage, fertility level, ER and AR transactivity (*loc. cit.*). Gynecology studies determined shorter duration of pregnancy and fetal growth, lower birth weight and shorter gestational age, longer menstrual cycles, deteriorating bone health and high chances of breast cancer (*loc. cit.*). All studies were from Nunavik or Greenland Inuits.

**Impacts on Arctic Wildlife Population**

Arctic animals are also affected by plastic pollution. Polar bears are top predators in the food chain and are known to be the most pollution-impacted species in the world, which is evident in the bio-tissues of polar bears (Kirk, 2016). Greenland is home to five seal species, among which is the harp seal, hooded seal and the ringed seal. The seals are an important part of Greenlandic hunting culture and an economic foundation in many families. When seals get entangled (with plastic pollution), it increases their energy consumption due to extra weight, ultimately leading them to consume food four times the normal amount (*loc. cit.*). Furthermore, they may be threatened by secondary ingestion of microplastics from fish which could later pose a threat to human health when consumed. Figure 26 is a conceptual map detailing how human interference by way of plastic pollution is subsequently impacting marine animal populations.

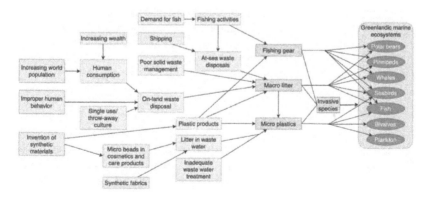

**Figure 26:** *Conceptual Map of Plastic Pollution impacts on Greenlandic marine ecosystems Adapted from (Kirk, 2016).*

Whales, which tend to ingest heavy amounts of microplastic while feeding, sometimes including fishing nets which can rupture their internal organs, thus giving rise to potential inconveniences towards humans when consumed as part of their traditional marine diet. Bivalves such as oysters and mussels have potential impacts on indigenous communities as they often eat the whole bivalve, including the digestive tract. In comparison, studies show that an average shellfish eater in Europe consumes up to 11,000 pieces of microplastic annually, which could lead to drastic impacts on their health (Kirk, 2016).

Due to lack of waste management in some of the Arctic coasts, the rate of recycling is low and the absence of incineration plants can result in open waste dumping sites that are close to the shore. This attracts polar bears and seabirds who mistake their contents as edible food and consequently ingest them. Another rising concern is the input of untreated wastewater into the ocean due to absence of wastewater treatment along coastlines (Kirk, 2016). However, installation of wastewater treatment plants in coastal areas can reduce the discharge of macro marine litter up to a certain extent (*loc. cit.*). The lives and values of the indigenous communities are at stake. A system could be put in place to avoid or reduce marine plastic debris that is currently ruining ecosystems in the Arctic.

### 3.4 Is There a Need for an Arctic Treaty?

After discussing in depth about how plastic pollution is affecting the Arctic Ocean from an environmental, social and economic perspective, the following section will discuss the importance of establishing a comprehensive political framework for the region able to exploit the monitoring measures that will be outlined in chapter 4. The ultimate goal is to support the decision-making process to design innovative solutions or enforce the already existing ones that will detect, monitor and mitigate marine plastic pollution.

There are two different parts of the Arctic's legal governance; the first is made up of a group of sovereign states, which have jurisdiction up to the lines delineating their borders. The second part comprises the central area free from any national sovereignty. Regulating national and international activities is important in the central (non-sovereign) area in particular. Therefore, the importance of having an overview of the political framework of the Arctic Ocean concerning plastic pollution is integral to implementing such an endeavor. Gaps within the current political framework can be filled per new agreements, but could also strengthen existing ones. Thus, the consideration of an Arctic Treaty is indeed warranted and will be expanded upon later in this report. For now, it is important to highlight the important organizational and political instruments currently in place from which an Arctic Treaty could be built upon.

### 3.4.1 The Big Two: UNCLOS and MARPOL

This overview begins with the general framework of international political instruments governing the Arctic. In 2008, a meeting in Greenland hosted the five coastal nations innervating the Arctic (i.e. Canada, Denmark, Norway, the Russian Federation and the United States); these nations expressed their commitment to the Convention on the Law of the Sea (UNCLOS), as an international instrument providing the essential elements to regulate, among other things, the protection of the marine environment (Dodds, 2013).

According to the Convention, the waters of the Arctic Ocean are considered international, with the exception of those that extend within the 12 miles (19.31 km) belonging to the continental shelf and Exclusive Economic Zones (EEZs) (United Nations, 1982). Coastal states, however, advocate for an interpretation of the law more favorable to their interests; they state that it is not possible to take the current coastline as a reference to delineate their boundaries, as the sea can move forward and backward with time (Vitale, 2010). Their solution would then be to extend the EEZ up to 350 miles (563 km), as opposed to the 200 miles (322 km) provided by UNCLOS (*loc. cit.*).

While such a claim may seem audacious on the legal side, it may also present environmental opportunities. Article 234 of UNCLOS, grants coastal states the ability to regulate at will the ice-covered areas within their national jurisdiction (Tufts University, 2021). This means they can take measures to control, prevent and reduce marine pollution in ice-covered EEZs (*loc. cit.*). If each of the nations had jurisdiction extending up to 350 miles (563 km) from their coasts, there would also be the ability to locally control the degree of pollution in their respective portions of the ocean within their legal purview, thus making monitoring and subsequent mitigation techniques easier.

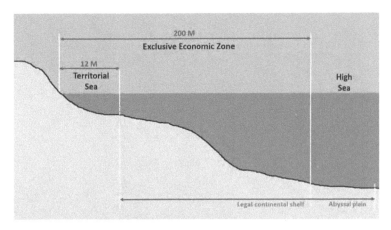

**Figure 27:** *Sea zones established by UNCLOS in 1982.*

Cleanup efforts have seen great success in the past; an excellent example translatable to the Arctic being the Antarctic. King George Island, home to many nations' Antarctic support stations, was fast becoming a junk pile as over a century's worth of refuse from explorers and abandoned mining interests built up (Marlow, 2009) and were not removed by the countries that simply saw the Antarctic as an exploitable resource. Under increasing international pressure, a concerted cleanup effort for King George Island Figure 28 started in 1992 and saw 1,500 tons of waste removed from the Russian Bellingshausen over the course of several years (*loc. cit.*).

The pollution crisis on King George Island was a significant motivator for the creation of the Protocol on Environmental Protection to the Antarctic Treaty 1991. The protocol banned all mining activities in the Antarctic and created the first regulations for handling waste in the Antarctic. A comparable Arctic Treaty with similar regulations would provide a legal framework of responsibility and accountability for nations and corporations operating or hoping to operate in the Arctic. This should be a pressing concern considering the past exploitation of Arctic lands as mining sites and nuclear weapons testing facilities, one of which is pictured left in Figure 29.

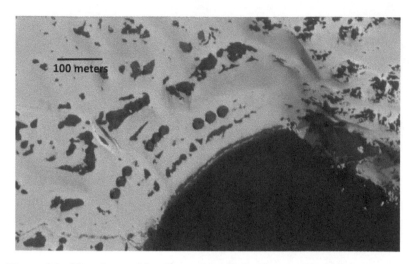

**Figure 28:** *King George Island experienced significant ecological damage from pollution on sites such as this one at 62°11' 33' S, 58°56' 07' W, just a short way north of Grikarov Point and Bellingshausen Station. Though the large storage tanks pictured here remain, a zealous cleanup effort through the 1990s removed 1500 tons of waste from this beach and the nearby base. Image taken from Google Earth*

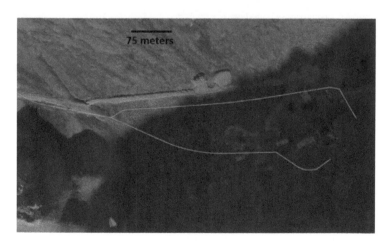

**Figure 29:** *Submerged mining site on Novaya Zemlya seen from space on Apple Maps near 73° 20' 57' N, 54° 40' 24' E. Dozens of such apparently abandoned sites can be seen less than 100 kilometers south of the location of the Tsar Bomba detonation. Image taken from Google Earth.*

Another very important international tool for the regulation of plastic pollution is MARPOL, imposing measures to prevent ships from throwing garbage into the ocean. In particular, Annex V, 'Prevention of Pollution by Garbage from Ships,' forbids the disposal of any kind of plastic into the water (IMO, 1988). All the Arctic States have signed the Convention. Additionally, in 2017, the International Maritime Organization also adopted the International Code for Ships Operating in Polar Waters, which is binding under MARPOL. The Polar Code also includes a section on environmental protection, which expressly prohibits the discharge of non-food waste into the water (Protection of the Arctic Marine Environment, 2021).

While UNCLOS regulates ice-covered areas only within the EEZ, MARPOL also covers the regulation of the high seas, with the possibility of categorizing some areas as 'Special' or 'Particularly Sensitive Areas.' One of these categorizations for the Arctic Ocean could be used for the Arctic seas as well, although this would involve a number of negotiations among all IMO members (Kirchner, 2017).

### 3.4.2 The Arctic Council: A Regional Effort

Given the unique characteristics of the Arctic Ocean, it has been necessary to create other regional mechanisms, since the international instruments, although undoubtedly useful to set the general binding framework, are not sufficient to implement a real strategy of targeted action.

The need for greater cooperation between the Arctic coastal nations (e.g. Sweden, Finland, and Iceland) led to the creation of the Arctic Council in 1996, which is a high level inter-governmental forum for collaboration, but without any rule-making power (Tufts University, 2021). Figure 30 summarises the modus operandi of the Arctic Council. These activities are run by six working groups, which include scientists, politics and other experts of various fields.

Agreements and cooperation     Data and knowledge     Monitoring     Assessments     Recommendations

**Figure 30:** *The five steps of the Arctic Council's working method.*

Most of the working groups are dedicated to activities which address marine plastic pollution mitigation. Activities have also intensified more since 2019, when Iceland, committed to combating this problem, took over chairmanship of the Council (until 2021). In particular, the Protection of the Arctic Marine Environment (PAME) working group is currently developing the Regional Action plan (RAP) on Marine Litter in the Arctic; additionally, the Arctic Contaminants Action Program (ACAP) working group is collaborating with states to improve onshore waste management, helping to reduce plastic waste in the water. AMAP documents the pathways and effects of pollutants, especially regarding the health of indigenous peoples, and one of its projects is developing Marine Litter Monitoring Guidelines (Arctic Council Secretariat, 2021).

As mentioned above, PAME started to develop the RAP on Marine Litter in 2017, a two-phase project collaboratively led by Canada; Denmark; Finland; Iceland; Norway; Sweden; the United States; the Aleut International Association (AIA), an NGO representing the Aleut indigenous people; and OSPAR, a mechanism through which fifteen states protect the Northeast Atlantic Ocean (Protection of the Arctic Marine Environment, 2021). The structure of the Plan is shown below (Figure 31).

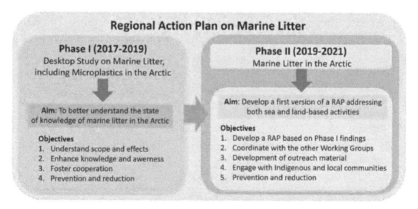

**Figure 31:** *Regional Action Plan on Marine Litter (RAP). As shown in the figure, the Plan includes two subsequent and interrelated phases, each of which has one aim and a list of objectives to be achieved by 2021. Adapted from (Protection of the Arctic Marine Environment, 2021).*

The RAP will represent the framework within which Arctic communities' efforts to address plastic pollution will be regulated. The gaps identified as a result of the working groups' assessments have served as a driving force for increased awareness and willingness to consider various mitigation strategies to come. Although RAP will not be completed until the end of 2021, Arctic communities are already largely engaged in developing the capacity building necessary to prevent the formation of a sixth gyre in the Arctic Ocean.

With the information gathered from the spatial and other monitoring discussed in the previous subsection and a broader awareness of environmental impacts of plastic pollution, policymakers have everything they need to outline a strategy for action. But will the Arctic Council be enough to shape the strategy into reality?

### The Structural Challenges

**Figure 32:** *Arctic Council structure. This figure shows all the different actors involved in the Arctic Council: there are 8 member states, 6 working groups, 6 indigenous organization and a plethora of observer states and organizations that are not involved within the decision making process (IMGBIN.com, 2021).*

In spite of its non-legal nature, the Arctic Council is a good example of peaceful cooperation between states, as it has never been dismantled by geopolitical tensions regarding international security (Exner-Pirot, et al, 2019). Moreover, its inability to make decisions that are binding on its member nations has not prevented it from having a productive role (Young, 2016), capable of directing and shaping the political agenda of Arctic states and beyond.

Despite being productive, the council's role does not allow it to oversee the implementation of national activities, thus making project monitoring difficult and enforcement impossible. Despite the creation of a tracking tool, the Amarok, it is still difficult to measure the progress of the various projects in a systematic way, mainly due to the lack of resources (Exner-Pirot, et al, 2019).

In fact, in addition to its legal nature, another obstacle to its effectiveness is the lack of regular funding from its members. Projects carried out by the council's various working groups are funded on an optional basis. All this makes it difficult to plan a long-term strategy, since projects are driven by funding and not vice versa (*loc. cit.*). An example of this is the efforts of Iceland, which, as the current chair of the Council, has been able to focus on combating plastic pollution. However, it is uncertain whether this commitment will remain in place in the years to come as other nations take the chairmanship, a role which rotates biennially.

**Enhancing the Arctic Council Effectiveness**
Despite its good example as a policy-shaper, the Council's effectiveness is undermined by legal limitations. In the 1990s, one of the conditions for the United States to join this intergovernmental body was that it was not to be created as a treaty-based international organization that would be binding on member states but rather as a forum that would operate on the basis of consensus (Exner-Pirot, et al, 2019). Consequently, there is nothing to suggest that the council's legal authority has become established after twenty-five years since its inception. While desirable, it is difficult to imagine that the Arctic Council could become an international organization with policy-making authority.

However, there are other elements that can be worked on, in order to make its work more effective. First, the strengthening of scientific cooperation, both among members and among working groups could be established. Also, as technological solutions, such as satellite monitoring, take on an increasingly important role in national strategies, scientific cooperation can act as a catalyst for greater integration of the Arctic Council. As Arctic states, 'institutional integration [can be seen] as a consequence of spillover effects of cooperation in one sector, creating a more comprehensive institution as a whole' (Binder, 2016, 128).

In this regard, it is very important to refer to the Agreement on Enhanced International Arctic Scientific Cooperation, a binding instrument signed by the eight member states of the Arctic Council in 2017, as a result of their desire to strengthen their scientific cooperation to jointly address major challenges in the region. Herein lies a concrete example of how science diplomacy can strengthen the political role of the Arctic Council (*loc. cit.*). Science, in fact, can be seen as a tool of soft power that can not only influence national policy decisions, but also change perceptions. More importantly, the current focus of Arctic states on issues such as plastic pollution is an example of this.

In this context, working groups assume the role of a bridge between science and policy. For this reason, it is important that the cooperation between them is further strengthened, perhaps through the adoption of a unified strategy to frame their action. In this way, it is possible to develop standards that are shared and accepted by all members, without the need for them to be binding.

Another way in which the Arctic Council could increase its effectiveness is by involving sub-national groups more closely, such as indigenous communities, but also regions and provinces. A bottom-up approach could in fact make its work more effective and efficient, as it would allow addressing problems by breaking them down locally into context-specific challenges. For this reason, Herrmann (2017) promotes the establishment of a working group on sub-national inclusion, which could consider more flexible solutions and create partnerships with the private sector more easily.

### 3.4.3 How To Solve This Fragmentation

To answer the initial question, 'Is there the need for an Arctic Treaty?', it would not be so easy today to obtain the adhesion of all the Arctic nations to sign and ratify a binding treaty; the non-ratification of the United States at UNCLOS exemplifies this. In addition, one of the conditions for their accession to the Arctic Council in 1996 was the prevention of such a council being bound by a treaty. It is therefore unrealistic to think that the Arctic states would want to bind themselves in an Arctic Treaty.

Nevertheless, an Arctic Treaty would indeed be effective at addressing Arctic plastic pollution. It may indeed be called for as the fragmentation of legislation regarding the Arctic Ocean between international binding instruments and regional soft power mechanisms may lead to increases in the claims of sovereignty by nations around in the central part of the Arctic (i.e. where no nation has sovereignty). This seems realistic when considering the aforementioned potential for increased trade routes through Arctic waters. In addition, the presence of a binding instrument would also allow for the creation of a body tasked with enforcing the law and enacting sanctions on those who do not comply.

However, the modern framework is not completely lacking. Both UNCLOS and MARPOL, though, do not specifically address the Arctic Ocean or contain binding provisions regarding pollution from ships at sea. For the remaining areas, as it happens in other nations of the world, national legislation governs where international law is lacking and it is precisely at this point that the Arctic Council is positioned. Indeed, it is possible to think of such a forum as a connection point between the global and the local, where both member states and sub-national groups can effectively cooperate with each other to find scalable and feasible solutions.

The diagram below (Figure 33) visually shows what the policy framework would look like once each actor is put in their place, calling out the importance of scientific cooperation, local inclusion and partnerships with the private sector.

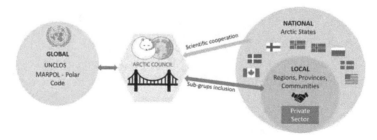

**Figure 33:** *The Arctic Council as a bridge between Global and Local policy. This diagram shows the role of the Arctic Council in connecting the international political framework that regulates plastic pollution in the oceans to both the Arctic States, through scientific cooperation and the local communities, including private sector, through a major sub-groups inclusion.*

To conclude, despite the similarity of the Arctic region to the Antarctic one, there are many differences at the legal level; for this reason it is difficult to think of an Arctic Treaty like the one existing for Antarctica. The presence of different nations and communities makes it difficult to agree on binding decisions. However, this diversity can be seen as an added value and a valuable tool when it comes to protecting the environment. In fact, the diverse needs of local communities can serve as a catalyst to prompt more immediate action to get nations to act quickly. The Arctic Council has great potential to become an effective mechanism that models, as a final step, global governance.

## 3.5 Chapter Summary

This chapter has provided an overview of how the Arctic Ocean is coping with the problem of a sixth gyre formation and the extent to which both indigenous peoples and the regional economy are suffering the effects of pollution. In particular, the rise of plastic in this remote ocean was first reported in the 1970s, adding further concern to this area of the world already suffering from the more ubiquitously known phenomenon of melting glaciers. The Earth's thermohaline circulation, in fact, moving plastic upward, also transports marine litter from the world's other oceans to the Arctic region.

The permanent effects of sea ice loss in the Arctic will create new economic opportunities in maritime activities by way of the creation of the TSR and expansion of the NSR and NWP. Global plastic pollution originates primarily from land based activities but evolving conditions in the Arctic may shift these estimates for the Arctic region to be more from marine sources. A correlation was identified between increased activities or usage of shipping routes and plastic pollution. These sources include more buoy dispersion, fishing nets, vessel paints, packaging, capsised containers and debris from collisions that damage or sink vessels. These sources are in addition to plastic coming from natural resource extraction, seaport activities and tourism boats. It is precisely these activities that, if unregulated, can increase pollution amounts. Being able to cut shipping times and fuel savings has already intrigued governments, such as Russia and China, who have already expressed their intention to use trade routes immediately, with a few economic centers already shifting upwards and displaying little respect for the environment.

With increased maritime activities, the increasing amount of plastic pollution puts at risk all those activities that can be carried out in the Arctic, which could be from the new trade routes on top of the abundance of natural resources, tourism and numerous seaports. At the same time, however, it is precisely these activities that, if unregulated, can increase the risk of pollution. In fact, one of the effects of global warming is the possibility of opening new shipping lanes that will undoubtedly intensify human activity, exacerbating existing environmental problems, including plastic pollution. This should not be ignored.

The other important consequence is that indigenous people living in the Arctic states are also indirectly affected by the problem, eating four times more seafood than the global average. Such seafood may ingest microplastics found in the ocean, which could lead to adverse health outcomes for both the sealife and the life forms, including humans, that eat it. This situation requires an intergovernmental regulation that only a joint effort based on cooperation can achieve. Therefore, the Arctic Council has been identified as the ideal mechanism to push states to do more, using in particular scientific diplomacy and the inclusion of subgroups. Although an Arctic Treaty is generally desirable, it remains more likely to work through soft power mechanisms as opposed to legal doctrine but should be supported by global and national binding legislation.

# 4

# Houston we have a Problem:
# Space Technology to Monitor Ocean Plastic

'It's surely our responsibility to do everything within our power to create a planet that provides a home not just for us, but for all life on Earth' - David Attenborough, *Naturalist/ Presenter.*

## 4.1 Introduction

In the previous chapter, it was determined how crucial it is to solve the problem of plastic that ends up in the ocean and breaks down into tiny pieces that are very harmful to life. However, it can be challenging to track plastic using conventional ground-based methods. Therefore the goal of this chapter is to provide an overview of technology capable of monitoring debris from space or near-space and use it to surveil the evolution of the amount of plastic in specific oceans to then apply mitigation measures such as removal by autonomous drones. Some studies, such as Biermann, et al. (2020), discuss the possibility of tracking the plastic using new and innovative methods with space, in this one it could be in the form of using optical satellite data to track and monitor the development of plastic patches in coastal areas using spectroscopy. Abrams (2019) focusses on a biomimetic walking robot concept to clean the ocean floors.

In that context, this chapter aims to present technologies with space or near-space components to detect, track and mitigate plastic debris with cutting-edge technology to get data, then serve as a first step to prepare for mitigation efforts while assessing the advantages and limitations of such technologies. Eventually, a clear determination of which is useful and applicable to the oceans will be presented, specifically in the Arctic context discussed in the case study.

First the issue of how to monitor currents in the oceans and water temperature to study the formation of gyres will be addressed. The focus then shifts to the issues of reflection and differentiating between plastic, ice and clouds automatically to monitor plastic from space. A discussion follows regarding new cutting-edge technologies such as underwater robots that communicate and navigate using satellites, concepts such as Zero 2 Infinity and World View balloons to detect plastic from high altitudes or space and using EO satellites for aggregated floating debris detection using Fourier Transform Spectroscopy (FTS) at various wavelengths. This chapter will close by discussing an integrated strategy to apply these technologies to the Arctic Ocean and its scalability to other oceans.

## 4.2 Why Are Space Technologies Important to Monitor Plastic Pollution?

Earth monitoring tools from space are accessible everywhere, often without needing a special computer or installing a specific software. Furthermore, it is possible to observe the whole planet if the right combination of satellites and orbits are chosen. The Copernicus Marine Service, for example, can be easily used, since it is an open data source and no additional program is needed on the utilised computer (European Space Agency, 2021a). This would allow savings in terms of money and resources. Moreover, EO satellites are already placed in orbit, and in this way it is possible to use the data they provide near real time, without waiting too long. Earth observation (EO) instruments allow to take very accurate images, thus resulting in the detection of plastic debris.

In some cases, as for the early detection of the sixth gyre that is forming in the Arctic Ocean, it would be useful to integrate other near-space technologies, such as the high altitude balloons this report will further explore, and in situ measurement in order to overcome space-based technologies limitations.

## 4.3 Assessment of General Requirements for the Monitoring and Tracking with Earth Observation Satellites

Ocean water contains different materials such as chlorophyll pigments, sediments and other suspended matter. These materials interact differently with photons depending on the wavelength.

This interaction allows water composition to be detected using optical remote sensing technologies (Dekker and Hestir, 2012). As discussed in chapter 2, European Chemicals Agency (2020) classifies marine plastic debris into macro-, meso-, micro- and nanoplastic. Due to its properties, plastic is sensitive to the photodegradation effect responsible for transforming macroplastics into microplastics. This decomposition also happens through abrasion. The positive argument is that suspended plastics in the water can be detected using EO satellites. This is important as this project aims to track and monitor macroplastics within the Arctic region in order to counteract the creation of another plastic debris gyre as well as prevent accumulations of macroplastics from sinking due to biofouling and ballasting, contributing to the formation of micro- and nanoplastics (Biermann, et al, 2020).

The first and most important requirement to the proposed solutions is that they should apply to every ocean. Therefore, the main idea is to enable tracking and monitoring macroplastics with EO satellites as it is possible to observe every ocean with them. In addition to this, and depending on the satellite, it is even possible to retrieve the observation data for free. Furthermore, the current COVID-19 situation underlines the importance of having a solution that works remotely as in situ solutions are currently hard to perform.

As shown in Figure 34, EO satellites have five primary requirements to fulfill. These requirements are the satellite's temporal, spatial, spectral capabilities, signal strength and orbit (Biermann, et al, 2020).

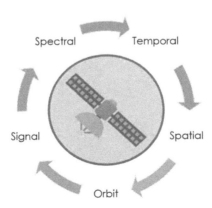

**Figure 34:** *The five requirements for Earth observation.*

Ideally, it would be possible to collect real-time information on marine waters to search for plastic debris. Unfortunately, the current state of satellite technology does not allow for realtime EO. Therefore, it is necessary to conduct a retrospective analysis of marine waters and approach real-time remote sensing as closely as possible (Hu, et al, 2015). Retrospective analysis means that the information gathered through remote sensing does not coincide with the actual water status. Another aspect of the temporal requirement is that the satellites must often fly over the analyzed location to establish a time series. As Dekker and Hestir (2012) explain, a time series is needed to determine the atmospheric and marine conditions that may occur daily, periodically and yearly. With the time series, it is possible to assess the condition, changes and trends to better predict the plastic debris movements. Therefore, the aim is to have a repetition time of a few days at most (Biermann, et al, 2020).

The EO satellite instruments' spatial resolution needs to be high enough to detect an accumulation of floating debris. The spatial resolution of the satellite depends on the type of plastic debris it should track. Since this project will focus on detecting macroplastics, only the spatial requirements for that type of plastic are presented. The detection of floating accumulations of macroplastic depends on the quality and amount of the reflectance as it needs to cover at least thirty to fifty per cent of the pixel to be noticeable.

Therefore, the satellite's spectral resolution depends on the aimed coverage (Biermann, et al, 2020; Topouzelis, Papakonstantinou and Garaba, 2019). The radiometric resolution is also essential as it provides information about the satellite's sensors' measured reflectance (Dekker and Hestir, 2012).

The Figure 35 from Dekker and Hestir (2012) shows the same image with four different spatial resolutions going from the high (upper left) to a low resolution (bottom right). The high spatial resolution allows for detecting a small accumulation of floating macroplastics through the visible spectrum (VIS).

In addition, consideration must be given to spatial sampling techniques, because it is not possible, in terms of energy and time, to sample all the oceans of the Earth for plastic waste at the same time. Since this is a geographical problem, it is recommended to study a particular area using the usual geostatistical approach of model-based design. As Dekker and Hestir (2012, 20) explain, 'in model-based sampling [...] the population is considered random and the sampling locations [...] are fixed'. As described in chapter 2, AI and machine learning are consequently required tools to analyze the data.

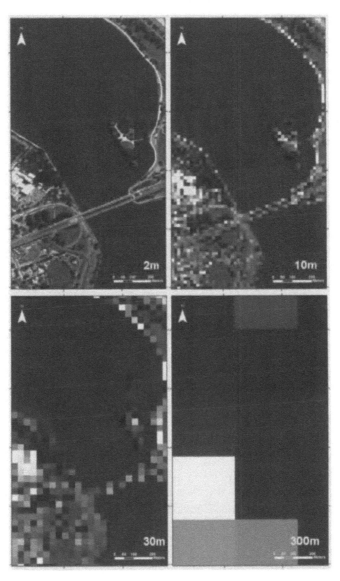

**Figure 35:** *SAR images with different spatial resolutions (Dekker and Hestir, 2012).*

Spectroscopic analysis gives the best results to detect plastic in the marine environment. The spectral resolution can distinguish plastic debris from natural sources of floating debris and seawater (Biermann, et al, 2020). The water absorbs NIR to SWIR, while plastic reflects light in the NIR region. Studies such as the one from Masoumi, Safavi and Khani (2012) suggest that plastic reflects in the wavelength range of 700-1900 μm. This study also shows that the reflectance is best in the spectrum of 1000-1700 μm. Effective EO satellites should therefore have a NIR spectrum of 1000-1700 pm. In addition to this, the satellites are also capable of detecting plastic accumulations through VIS imaging. Ergo, these spectra (NIR-VIS) are demanded for the spectral analysis techniques described in Section 4.3. (Topouzelis, Papakonstantinou and Garaba, 2019).

In addition to the spectral resolution, consideration must be given to the atmosphere, the effects of which need to be corrected on the sensing satellite (Zhu, et al, 2019). The weather and atmospheric conditions, such as clouds and ionospheric scintillation, impair signal strength and detection. For that purpose, correction algorithms, like Sen2Cor and ACOLITE, need to be used to counter these kinds of effects (Dekker and Hestir, 2012). The detection and removal of cloud effects are also necessary and are described in more detail in section 4.6.2.

The last requirement to consider for EO satellites is the orbital parameters. The chosen orbit determines the coverage of the satellite. Moreover, the approach is based on passive remote sensing. The sensors rely on the reflected sunlight, therefore, a Sun-synchronous orbit is required. A Sun-synchronous polar orbit is the best option for remote sensing of the non-polar oceans. They are regrettably not suitable for coverage of high latitudes due to their low elevation angles. Since this project focuses on the Arctic Ocean, coverage of the Polar regions is a requirement for the recommendation. Trishchenko and Garand (2011) suggest in that case to use a highly elliptical Molniya type orbit (HEO). They add that it is possible to continuously cover the entire Arctic region (58° 90°N) with only two coplanar satellites. They also recommended that the apogees are at 95°W, 5°W, 85°E, 175°E.

## 4.4 Monitoring Ocean Current Tracking

### 4.4.1 Water Temperature

Water temperature is a property that is typically used only in reference to the thermal energy of the water. However, knowing the temperature of the water can aid in determining a number of other characteristics, such as density, pH levels, conductivity, salinity, dissolved gas concentrations and many other important characteristics.

Oceans cover roughly seventy-one per cent of the Earth's surface. Over extended time periods, the oceans can retain more absorbed solar heat than the atmosphere or the land (ESA, 2021). The prevailing ocean currents are driven primarily by differing water densities, winds and tides. Differences in water density are typically driven by differences in temperatures, salinity and other dissolved solids (NOAA, 2011). As the Earth rotates, the ocean's waters and winds get diverted from a straight-line path. This causes ocean currents to curve right in the northern hemisphere and similarly left in the southern hemisphere (*loc. cit.*). Water temperature is very influential in the global climate. Sea surface temperature is a parameter used to predict the weather and atmospheric model simulations and also important for studying marine ecosystems (NOAA, 2021). Like thermometers in the atmosphere, surface temperatures can be measured using sensors from satellites which can be found in Table 4.

**Table 4:** *List of satellites used to measure sea surface temperature.*

|    | Satellite | Sensor used | Operator | Reference |
|----|-----------|-------------|----------|-----------|
| 1  | Terra | MODIS | NASA | (NASA, 2021b) |
| 2  | Aqua | MODIS | NASA | (NASA, 2021a) |
| 3  | Terra | ASTER | NASA/JAXA | (NASA, 2021b) |
| 4  | POES | AVHRR | NOAA/MetOp | (OSPA NOAA, 2020) |
| 5  | Sentinel-3 | SLSTR | ESA | (ESA Sentinel, 2021) |
| 6  | Landsat 4, 5 | TM | NASA | (eoPortal Directory, 2012) |
| 7  | Landsat 7 | ETM+ | NASA | (Masek, 2021a) |
| 8  | Landsat 8 | TIRS | NASA | (Masek, 2021b; USGS, 2021) |
| 9  | Sentinel-6 | GNSS-RO | EUMETSAT/NASA | (eoPortal Directory, 2021) |
| 10 | ERS 1 and 2 | ATSR | ESA | (ESA, 2021a) |
| 11 | Envisat | AATSR | ESA | (ESA Earth Online, 2021) |

**El Niño and La Niña**

The detection of temperature fluctuations reflecting the changes in oceanic cycles each year is an important forecasting tool for regions around the world. El Nino causes an upwelling of deep Pacific water near South America, depleting the nutrients that fish require to survive (NOAA, 2020). The subsequent loss of harvestable seafood stresses nearby fisheries and potentially forces their fleets further into the Pacific or, potentially, Atlantic Oceans; this is a potential means for changing the significant macroplastic inputs due to fishing in multiple gyres. El Nino patterns also cause significant downpours of rain in Ecuador, Peru and southern Brazil but correlates to low precipitation and droughts across North America, Africa, southern Asia and Australia (American Geosciences, 2016).

El Nino and La Niña are greatly influenced by background seasonal changes. They show major variations in their spatial structure and seasonal evolution. These cycles cause periodic changes in the Pacific Ocean surface temperatures, which impacts the global weather (American Geosciences, 2016). These cycles last for 9-12 months and occur every three to seven years (*loc. cit.*). During El Nino, temperatures near the equatorial region are hotter and dryer due to the equatorial waters flowing along the surface of the oceans (NOAA, 2021). Under El Nino conditions the westerly winds are weaker and permit more precipitation in much of South America (American Geosciences, 2016). During La Niña, temperatures near the equatorial region are colder and wetter than normal (*loc. cit.*). In this scenario, the east-west trade winds strengthen and the warm surface waters are pushed west across the Pacific (American Geosciences, 2016). A visualization from the World Meteorological Organization (WMO) below in Figure 36 shows the trends of the currents with respect to their oceanic sources.

Morishige, et al. (2007), found during their study on marine debris deposition at French Frigate Shoals that the plastic debris deposited during La Niña events was significantly less than during El Nino or non-event periods. The interactions between the El Nino currents and the surface winds appeared to be responsible for a significant accumulation of debris on the north-western Hawaiian Islands. Water temperature and salinity are responsible for differences in density and these differences are the driving force of a part of the ocean circulation known as thermohaline circulation (Rahmstorf, 2006). Due to the

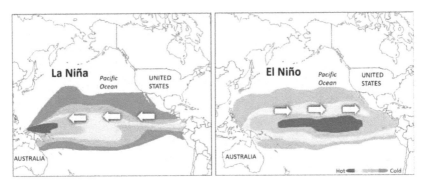

**Figure 36:** *Trends of the currents with respect to their oceanic sources. Adapted from (Anwar, 2017).*

rise in global temperatures, thermohaline circulation causes marine plastic debris to move into the Arctic region. For more information regarding thermohaline circulation, refer to section .

According to Howell, et al. (2012) and Maximenko, et al. (2019), El Niño and La Niña are important to understanding the circulation of plastics in the subtropical gyres, although little evidence was found to confirm a direct impact. The relationship between seasonal events and plastic debris accumulation patterns could contribute towards the fluctuations in plastic accumulation in gyres, the rate at which plastics degrade and facilitate the transportation of debris to other oceans. As determined in section 3.2.4, plastic debris from subtropical gyres by way of Earth's ocean circulation mechanisms brings plastics into the Arctic Ocean. With anthropogenic climate change, the intensity of these events is increasing and could potentially impact the rate at which plastics enter the Arctic.

### 4.4.2 Formation of Gyres

#### The Origin and Route of the Plastic Debris That Can Be Found in Oceans

EO can be used to monitor plastic debris routes using remote sensing applications. As previously mentioned, most plastic sinks or rests a small distance below the surface and thus cannot be effectively tracked from the orbit. Floating debris, however, can be seen from orbit (Martínez-Vicente, et al, 2019). Plastics in the ocean originate from two sources: terrestrial sources which account for between 4.8 and 12.7 million tons of plastic and the river source which represents around two million tons (Lebreton, et al, 2017).

Plastic routes are still under thorough investigation today because tracking plastic as it moved was not a feasible focus for study historically. The greater part of plastics in the ocean actually come from human activities on the continent (Li, Tse and Fok, 2016). Much of it flows from rivers into the oceans, with the top 20 polluted rivers accounting for more than two-thirds of all rivers in the world (Lebreton, et al, 2017). As described in section 3.3.2, plastic debris also originates from marine vessels.

**Formation of Ocean Gyres Through Water Currents**

Microplastics entering the ocean come from a variety of sources, as shown in section 2.2.2, and are transported from coast to sea. These currents are the result of wind and atmospheric movements and are responsible for the creation of garbage gyres. As of the writing of this paper there exist five: The North Pacific Gyre, South Pacific Gyre, North Atlantic Gyre, South Atlantic Gyre and Indian Ocean Gyre (5Gyres.Org, 2020). It is a result of the opposite current that interacts with each other, producing circular currents responsible for trapping floating debris and aggregating marine litter to form these garbage patches. Using scientific buoys operated by NOAA and working alongside NASA, it has been demonstrated that once buoys are drawn into the gyres they remain in circulation until they break down into microplastics (Shirah and Mithchell, 2015).

Mathematical modeling of ocean movements is already being used to determine how such gyres actually formed. It is based on a few assumptions. The first one is that the ocean is considered as incompressible non-viscous fluid where the density is homogenous and the pressure is also considered in a perfectly balanced state with gravitational force described as $\delta p/\delta z = -\rho \times g$ (Saint-Raymond, 2010). The last approximation is about describing the movement only horizontally. Complex equations can describe it using Saint-Venant equations as well as the Coriolis force. The Coriolis force is an inertial force that is perpendicular to the movement of a corps in rotational referential and creates a curved trajectory (Figure 37). Due to the Earth's rotation, the Coriolis force deflects any movement to the left direction in the Southern Hemisphere and to the right direction in the Northern Hemisphere.

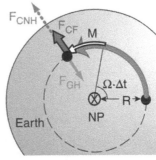

**Figure 37:** *On the left is what we can observe as the results of the Coriolis Force. On the right is a diagram showing all the force that applies to the system and the resulting trajectory with the North pole as the referential (Altendorf, 2020; Lehning, 2018). Here R means radius of curvature, NP = North Pole, $\Omega.\Delta t$ - angular movement. $F_{cf}$ is the centrifugal force, $F_{GH}$ and $F_{CNH}$ are the gravity force and centrifugal force in the local horizontal respectively.*

Heat transportation is another factor to consider as cold wind and water flow from the poles and warmer wind and water from tropical areas flow away from the equator. This atmospheric and oceanic circulation fuels the gyres' rotations. These gyres are responsible for water circulation throughout the globe and regulate certain factors such as temperature, salinity and nutrient flow.

This kind of ocean circulation type is called the wind-driven circulation (Wells, 2015) and is the strongest of the two types, especially on the surface. This mechanism is shared by all ocean gyres because the cold and hot wind from the poles and the equator respectively interact with the surface water. Wind impact on the water will exert a force on the water 90° to the right side if it is from the north pole and to the left side if it comes from the south pole and going downward in a circle (as displayed in Figure 38) to the depth of the ocean for less than 100 m. This phenomenon is called Ekman pumping or transport. This is the reason why plastics do not simply aggregate at the surface and actually sink and stack, creating an extremely thick layer of plastic similar to an island.

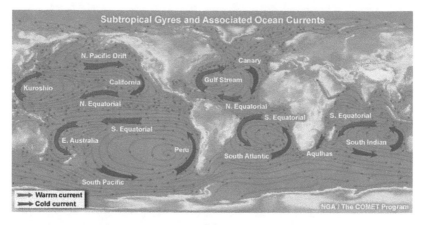

**Figure 38:** *Subtropical gyres and associated ocean currents as a result of hot wind coming from the equator and cold ones from the poles (Cooke, 2016).*

As the current is going more in-depth, it is affected by the hydrostatic pressure that is obtained using $p = g \times p \times z$, where z is the depth, $p$ is the density that changes with the temperature and depth, and g is the gravity acceleration. Horizontal differences in the salt concentration and the temperature are called the thermohaline circulation which is the second type of ocean circulation. Therefore, it is necessary to adapt computer modeling because the change of density layer can affect the trajectory and speed of such a current (Bogden and Edwards, 2001). Something else to consider is the influence of the temperature of the water impacted by either hot or cold wind on the real density of each layer and thus both of those types of ocean circulation are actually dependent on one another.

## 4.5 EO Satellites Utilization for Aggregated Floating Debris Detection Using Fourier-Transform Spectroscopy at Various Wavelengths

Satellite remote sensing has demonstrated its efficiency in plastic detection. The latest research shows that an accuracy of eighty-six per cent can be reached for the detection of plastic patches around 10x10 m. Indeed, it is the only technique currently available that is capable of providing great quality imaging and continuous ocean monitoring. Not so long ago, it was not possible to do so because of low-resolution satellites and periods of days even between two shots. However, thanks to EO satellites from commercial services and from the ESA, it is now possible to achieve a greater resolution with several spectral bands with a revisiting period of a few hours which is already perfect. The cost of such an operation is also quite low because most of the data that is possible to get from EO satellites is free and the diversity of data collection is quite significant. This is very important because it can differentiate natural marine litter from macroplastics (Biermann, et al, 2020). Plastics will aggregate in the ocean as discussed in section 4.3. and create plastic gyres. Only the plastic that is on the surface can be detected using space technologies because the light does not reach the bottom of the ocean. Both of those types are macroplastics that can be monitored until fragments start sinking. As macroplastics remain in the water over time, they start decomposing into microplastics. This action is accelerated by photodegradation which makes them fragment into pieces under ultraviolet radiation as explained in section 2.4.1. The degradation makes their detection harder, meaning other more complicated approaches are therefore needed (Klein, et al, 2018). There are studies that assume that microplastic prediction can be done by the analysis of water turbidity but at the moment this assumption has not been proven yet. For this reason this paper focuses on macroplastics because conventional tracking approaches are practical solutions to the reduction of those debris and will reduce microplastics as a result.

To track macroplastics, the best way is to use existing technologies that use Fourier-transform spectroscopy and more particularly the reflection method as mentioned in section 2.4.1. This type of spectroscopy modifies the wavelength received into another wavelength and the reflection method can be used at different wavelengths. The first technology is the NIR and SWIR that peak at those wavelengths; the second is optical and visible imaging that with a sufficient resolution could be a great solution for plastic detection; the third is radio wave imaging using a Synthetic Aperture Radar (SAR) that investigates the surface roughness of the water surface in the C-band.

### 4.5.1 Infra-red Imaging of Floating Debris Aggregation Using Space Technology

Using a mathematical approach to determine precise light reflectance from the oceans, it is possible to detect plastic in water. Using NIR and SWIR spectroscopy from space and airborne detectors it is possible to track marine plastic debris. The mechanism of plastic detection relies on searching for specific patterns of plastic absorbance, which takes place in the NIR spectrum. A typical pattern of plastic reflectivity properties with respect to the properties of grass, saline soils, water and sandstone can be found below:

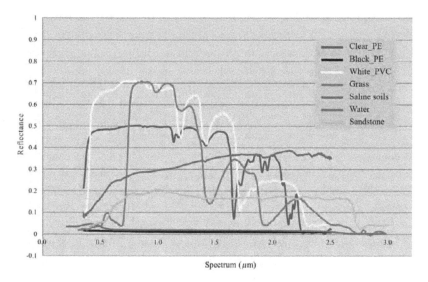

**Figure 39:** *A typical absorbance spectrum of some types of plastics (Lu, Hang and Di, 2015).*

From this figure, it can be seen that different kinds of plastic have a very steep decrease in reflectance after approximately 1.7 μm in wavelength. It can also be concluded that black plastics are very hard to track due to their high absorbance. This fact makes analysis complicated, as the share of the dark plastic litter in some regions may reach as high as sixty per cent (Migwi, Ogunah and Kiratu, 2020).

One of the problems that arises in analyzing these templates is distinguishing them from vegetation. Luckily, with regard to the oceans, this can be successfully done by the NDVI, commonly used to estimate the amount of green vegetation. Nevertheless, apart from the vegetation, other objects may exist. The possibility to distinguish between the different objects on the water surface was profoundly considered by Biermann, et al. (2020). In their work they have analyzed different reflectance patterns of the potential objects on the sea surface. Besides plastic, these objects included seaweed, timber, seafoam and pumice. They concluded that some of these objects may be grouped by a correlation with an additional index. Hence, they have introduced a new metric - the FDI, an empiric metric for debris identification and have managed to cluster these objects against it in order to aggregate them by type. After this, a machine learning program was used to analyze and standardised data.

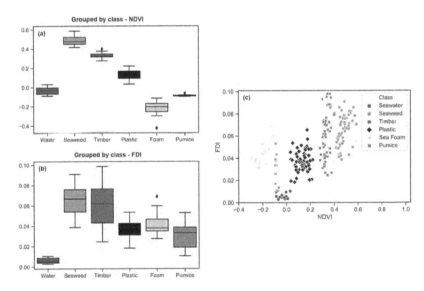

**Figure 40:** *An example of clusterization of different types of objects on a sea surface. Those Index are used to differentiate plastic from natural marine element such as water and vegetal debris (Biermann, et al, 2020).*

As a development of this idea, based on this experiment, Themistocleous, et al. (2020) have introduced another metric: the Plastic Index (PI). By deploying artificial plastic targets that were used as representative items of plastic pollution in a sea, and observing this area via Sentinel-2, they processed the data and compared the Sensitivity Analysis Value for all potential metrics, which could be used for plastic detection. The PI metric tends to be the most efficient, giving a strong spike on the plot, hence showing an even more pronounced correlation with plastic litter. The sensitivity analysis value shows what margin in plastic detection takes place for any given index in comparison with other indices. The most significant result of this study was that subpixel plastic clusters can be detected.

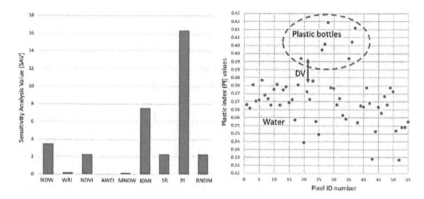

**Figure 41:** *The comparative analysis of the different metrics used for plastics identification (Themistocleous, et al, 2020). A strong spike of the Plastic Index is seen in comparison with other indices. This shows a promising potential of PI usage for plastics detection.*

Other experiments have been performed. Topouzelis, Papakonstantinou and Garaba (2019), also created target material using different plastic items. One of the three targets was composed of aggregated empty bottles, the second one of plastic bags and the last one was made of nylon fishing nets halfsubmerged in a 100 square meter support. This support would prevent plastic from sinking and place at least thirty meters from the beach and left in the sea for only one day that would be seen from orbit by Sentinel-2A and using airborne craft. The satellite data was acquired and then analyzed by the ESA Software which used deep water as the minimum reflectance in comparison with the target. As the resolution was still low (about 10 meters for Sentinel-2) target coverage in each pixel had to be determined and was therefore extracted.

Bottles and plastic bags will reflect light in the NIR with this resolution and if more than fifty per cent of the pixel is covered by water, the reflectance does not become significant (Topouzelis, Papakonstantinou and Garaba, 2019). For more than thirty per cent of plastic in one pixel however, values from the NIR spectrum will increase. Therefore, this will confirm that there is plastic in this location. This data shows that it is quite efficient to use satellite imaging to identify plastic debris aggregation in the sea. ESA's Sentinel-2 satellite NIR detector is quite efficient already to detect aggregated floating debris in the oceans. Monitoring marine debris aggregation and its routes yields the opportunity to understand patterns and how to slow the formation of gyres or to stop it entirely. This could also be done by drones or unmanned aircraft systems and space technologies with optical detectors.

### 4.5.2 Optical Imaging of Floating Debris Aggregation Using Space Technologies and Airborne Platform

Regarding the visible spectrum, the reflectance signal for white plastics was high because it does not absorb any significant light but the colored plastics have a maximum of reflectance at a wavelength corresponding to their color, as shown by Garaba, Aitken, et al. (2018). Their experiment was an overall success despite the resolution using Sentinel-2 optical detectors for at least two targets: The bottles and plastic bags. However, the fishing net target was not very well detected because of the dominant water background that reduced plastic optical reflectance. Transparent and blue plastics may also be a problem as they will let the reflectance of water be detected by the satellite. To this day Sentinel-2 only has three optical detectors which are red (665nm), green (560nm) and blue (490nm). This data could be supplemented by other EO satellites that have other detectors in the visible spectrum or simply by using the IR or SWIR spectrum detectors from Sentinel-2 to confirm it.

Utilizing an airborne strategy to get a more precise identification of small patches seems to be the solution for local and exceptional occasions since the resolution is around five centimeters. However, the flight time of such a craft is low (less than one hour) and therefore cannot be utilised all the time even in a high number for cost and power reasons (Topouzelis, Papakonstantinou and Garaba, 2019).

However, helium balloons can stay in the air for up to thirty days, as discussed in section 4.7.2 and these can be equipped with any detector with a better precision than satellites.

Optical and infra-red imagery are without a doubt an elegant solution but are dependent on two factors. The first one is that satellites and drones alike do not possess their own source of illumination. The second is that weather conditions can blind the satellites and prevent UAV's from flying safely. Radar imagery has been used to monitor the oceans because it possesses its own illumination source and penetrates clouds and most weather conditions (Themistocleous, et al, 2020).

Another factor to consider is the long revisiting period for the two Sentinel-2 satellites. By design, it covers primarily the equatorial plane and every ten days views the poles (Figure 42). It is also important to note that large plastic patches that can be detected with EO satellites such as Sentinel-2 are to a significant degree formed by the oceans' gyres, which are often not surveilled by optical satellites.

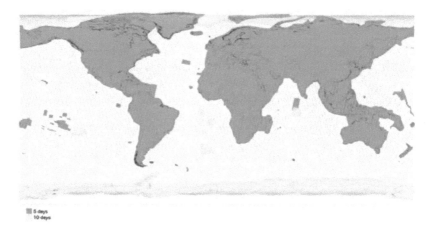

**Figure 42:** *Coverage area of Sentinel-2 shows the constraints of its usage to monitor oceans. It can be seen that the temporal resolution is quite low for a significant portion of the ocean. The color corresponds to the time between two revisits, 5 days in green and 10 days in yellow. (ESA, 2021)*

### 4.5.3 Radar Imaging of Floating Debris Aggregation Space Using Space Technology

The satellites with SAR are considered the next promising step in plastic detection. The first successful experiments that utilised SAR to find obstacles on the surface were conducted in 2018 by Topouzelis, Papakonstantinou and Garaba (2019). During this work, they analyzed the difference in the backscatter pattern of the litter on the water surface. They managed to find the strict difference between water and litter. Apart from the ability to penetrate through clouds, the main advantage of this method is that it can detect only physical obstacles on the surface. Therefore, additional calculation of different indexes, clustering and algorithms of classification is not needed. The other advantage of SAR systems is better spatial resolution and different analysis modes. For Sentinel-1, a practical resolution of five by five meters can be achieved in a strip map mode (ESA, 2021), which is sufficient to detect relatively small plastic patches. The main disadvantage of SARs with respect to plastic detection is the inability to tell the patches from ships. Both have similar dimensions and may have similar velocities on the surface. This makes SAR a good augmentation to optical analysis.

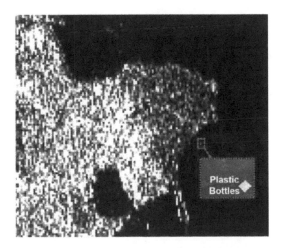

**Figure 43:** *An example of visibility of plastic debris on a SAR image taken by Sentinel-1 (ESA) with the white and grey dots displaying the land and a few grey pixels that display the plastic bottle target surrounded in red. This target is in the middle of the sea shown by the black background (Topouzelis, Papakonstantinou and Garaba, 2019).*

The two Sentinel-1 Satellites use SAR (Figure 43), and unlike Sentinel-2 satellites, they are in a polar orbit which is very useful for the aggregated floating debris located in the Arctic ocean known as the sixth gyre. Since SAR will detect the difference between two heights, it is very useful to detect small patches because the plastic does not only aggregate on the same horizontal plane but vertically as well. One could argue that the resolution of Sentinel-1 is not sufficient to track all the new patches in the ocean but currently Sentinel-6 is available and displays a much better resolution.

## 4.6  How Do We Monitor Plastics from Space on an Ice-covered Ocean Surface?

This section will discuss three major challenges faced to monitor and track plastic pollution in the Arctic Ocean: sea ice, clouds and satellite signals. It is important that these challenges are considered before proposing any technical space-based solutions to tackle the growing problem of plastic pollution and to understand what technical requirements are needed to overcome these challenges. These challenges are a few of the main reasons the Arctic remains under-studied.

### 4.6.1  Ice and Satellite Signals

The biggest challenge of monitoring and tracking plastic pollution is the existence of sea ice. Sea ice refers to frozen seawater floating on the ocean's surface that forms, grows and melts in the ocean (National Snow and Ice Data Center, 2020a); therefore, it does not include glaciers, ice shelves, icebergs, or ice caps because they originate on land. The thickness of Arctic sea ice varies tremendously throughout the year but on average ice is two to three meters thick, with some Arctic regions four to five meters thick (National Snow and Ice Data Center, 2020b) and an overall asymmetric distribution. This means that current EO methods cannot properly detect plastics under the sea ice surface, thus requiring measurements to be taken during warmer months or measurement techniques to be conducted using autonomous underwater vessels (AUVs). In Figure 44 are examples of micro- and macroplastics found in the sea ice which illustrate the plastic problem under sea ice conditions.

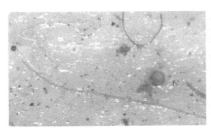

**Figure 44:** *Plastics Frozen in Arctic Sea Ice. Shown (left) are microplastics that have become embedded in the ice and (right) are macroplastics caught in water as it freezes (Hood, 2019; Quinn, 2019).*

Considering that Arctic sea ice can be as thick as four to five meters, there are also limitations in the effectiveness of satellite signals to communicate with vehicles under the ice. Already underwater environments pose challenges to communication, localization, and navigation of AUVs due to complex water properties. With the added complexity of sea ice, relying on radio communications and GPS systems is insufficient (González-García, et al, 2020).

In order to overcome these challenges, new and existing technologies must be adapted to monitor and track plastic within the sea ice and through the cloud cover. As mentioned in the previous section under assessment of requirements, it is important to keep in mind the temporal, spatial and spectral requirements when choosing an appropriate system to track plastic trapped within sea ice.

Sea ice covers large areas of the Arctic and the most feasible way to get an overview of the territory is to view it from space. The integration of space/air borne with in-situ technologies can be adopted to deal with plastic trapped in sea ice. As addressed in previous sections, underwater robots can be coupled with passive microwave technology for this detection, since objects at the Earth's surface emit both infrared radiation and microwaves at low energy levels, and clouds are known to emit lower microwave radiation compared to sea ice. This makes it easier for microwaves to penetrate clouds and detect plastic within sea ice during the day and night, regardless of cloud cover (National Snow and Ice Data Center, 2020a).

A second approach could be a combination of a mathematical model called Imaging FTIS and an automated polymer identification approach that can yield high resolution chemical images followed by sea ice coring using a Kovacs nine cm diameter corer (Peeken, et al, 2018). A third approach could use satellite altimetry with machine learning algorithms such as decision trees and random forest to estimate plastic debris in Arctic sea ice (S. Lee, et al, 2016).

Alternatively, tracking and monitoring plastics are possible by modeling the microplastics trapped during sea ice formation and drift and then tracking the release of the plastic during melting times. Sea ice acts as a medium for the redistribution of contaminants. Specific locations of interest include the shallow Siberian shelves located in the Eurasian Basin and the Beaufort Sea in the Amerasian Basin. Obtaining figures on microplastics entrapped during formation could be representative of microplastics present throughout the ocean.

### 4.6.2 Clouds: How to Differentiate Clouds Automatically?

Another significant challenge is the cloud coverage over the Arctic that makes it more difficult to acquire measurements on the ocean's surface which will only become more difficult as ice melts. He, et al. (2019) found that based on CALIPSO satellite observations and measurements as sea ice retreats cloud-coverage over ice-free regions is fixed at about eighty-one per cent, further reducing the albedo effect feedback as the light reflects back down and melts even more sea ice surface. This makes it important to be able to differentiate clouds from other sources of reflectance.

A cloud is a visible mass of concentrated water vapor in the atmosphere, which generally floats far above the ground. Various cloud types are named based on their composition and how high they float in the troposphere (Puiu, 2017). There are different types of clouds based on altitude and their unique characteristics (University Corporation for Atmospheric Research, 2019).

**Table 5:** *Different types of clouds (University Corporation for Atmospheric Research, 2019).*

| Main cloud type | Description | Clouds | Altitude |
|---|---|---|---|
| High-level clouds | Consists of ice crystals, they are mostly thin, sneaky, and white due to cold temperatures in the troposphere. | Cirrocumulus, cirrus, and cirrostratus | 5 km-13 km |

| Mid-level clouds | Consists of liquid water droplets, ice crystals or a mixture of the two, including supercooled droplets (i.e. liquid droplets whose temperatures are below freezing) depending on altitude, temperature in the troposphere. | Altocumulus, altostratus, and nimbostratus | 2 km-17 km |
|---|---|---|---|
| Low-level clouds | Consists of liquid water droplets, ice crystals or a mixture of the two, including supercooled droplets (i.e. liquid droplets whose temperatures are below freezing) depending on altitude, temperature in the troposphere. | Stratus, nimbostratus, and stratocumulus | 0 km-2 km |
| Clouds with vertical growth | Clouds that expand upward rather than scatter across the sky. | Cumulus, and cumulonimbus | 5 km-13 km |
| Unusual clouds | Clouds that are formed uniquely and are not grouped by height. | Lenticular, mammatus, and Kelvin-Helmholtz | Any |

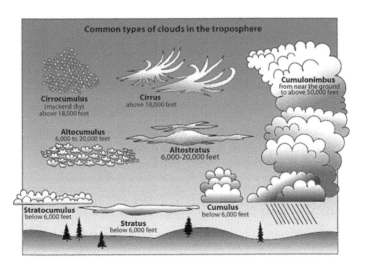

**Figure 45:** *The Figure shows common types of clouds that can be found in the atmosphere (University Corporation for Atmospheric Research, 2019).*

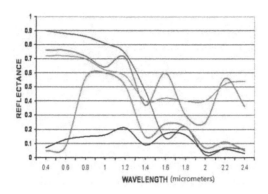

**Figure 46:** *Typical reflectance values for snow (blue), bare soil (black), forest canopy (pink), and cirrus (red) and stratus (green) clouds as a function of wavelength in μm (Jedlovec, 2009).*

Researchers employ cloud detection techniques using cloud physical parameters such as shape attributes, fusion of multi-scale convolutional features of cloud net, color transformation, cloud density, cloud shadow, clear-sky background difference and feature extraction of the image (Mahajan and Fataniya, 2020). Clouds at visible wavelength have a high solar reflectivity in comparison with most surface characteristics.

**Table 6:** *Cloud detection methods using classical algorithm approach (Mahajan and Fataniya, 2020).*

| Cloud forms | Satellite/sensor used | Parameters for cloud detection | Accuracy |
|---|---|---|---|
| Cloud/no cloud | NOAA images, GOES 12, MODIS, VIRR | Application of color models (HSV AND RGB COLOR MODEL) | 87.6%, 91.1%, 92% |
| Cloud/no cloud | Hyperspectral data from the hyperion EO-1 sensor | Histogram, T-ESAM, MRF model, dynamic stochastic resonance (DSR) model | 96.28% |
| Cloud/no cloud | The NASA's Aqua satellites, (INSAT-3D) | FSI-Fog Stability Index (surface temp, point temp & wind speed) | 84.63% |
| Cloud cover detection | NOAA's GOES-13 satellite image of 3600 × 3000 res | Adaptive thresholding-based approach | 89% |

| Cloud region | GOES-16 ABI, PSU WRF EnKF dataset | Gradient-based algorithm (morphological image gradient magnitudes) | 98% |
|---|---|---|---|
| Cloud/no cloud | Kalpana-1 and INSAT satellites | INSAT cloud masl algorithm, BT threshold test | 84% |
| Cloud/no cloud | ZY 3 satellite images | Image feature (gray scale vector, frequency, texture) | 90% |
| Cloud/no cloud | NOAA-AVHRR Satellite Data | Bayesian classifier, regression | 97%, 80% |
| Cloud/no cloud | Sentinel-2 Level 1C | Fmask algorithm (Cloud Displacement Index) CDI | 95% |
| Cloud/no cloud | GaoFen-5 Satellite | Dynamic threshold and radiative transfer models | 95.07% |
| Cloud/no cloud | MODIS DATA | Multi-spectral syntheis method-cloud detection | 95% |
| Cloud/no cloud | GF-1 and IKONOS | Spectral Indices method cloud index (CI) and cloud shadow index (CSI) | 98.52%, 89.12% |
| Ice/snow cloud | Landsat 8 images | Fmask and ACCA, algorithm function of Fmask algorithm, ACCA | 92.09% |
| Cloud/cloud shadow | GaoFen-1 data | Multilevel features fused segmentation network | 98.69%, 98.92% |
| Cloud/no cloud | 200 HJ-1/CDD and GF-1/WFV images | Automatic cloud cover assessment (ACCA) | 90% |

There are various types of cloud detection, including cloud/no cloud, thin cloud/thick cloud, snow/ice detection and cloud/cloud shadow. Radar imaging sensors and infrared imaging sensors can be used to track and monitor plastic on ocean and sea surfaces from space under cloud cover. For infrared imagery and radar imagery, refer to sections 4.5.1 and 4.5.3. Land or ocean surface features are difficult to detect using optical sensors under cloud cover. However, the same functionality can be monitored under zero or very low cloud cover.

### The Two Methods of Cloud Detection Approach

**Classical algorithm-based approach:** The classical algorithm method follows known specific steps for specific image input and the output will vary depending on the input image and algorithm used. Many cloud detection methods are based on the threshold but have poor universality (Mahajan and Fataniya, 2020). There are different methods using the classical algorithm approach and each method has its known specific steps for the input. These methods are given in Table 6.

**Machine learning approach:** Machine learning is a part of AI which allows systems to automatically learn depending on existing data. Machine learning techniques are extremely flexible and less complex because they simulate training decisions but are inconsistent because model learning relies on input data (Mahajan and Fataniya, 2020). The different methods using the machine learning approach are given in Table 7.

**Table 7:** *Cloud detection methods using machine learning approach (Mahajan and Fataniya, 2020).*

| Cloud forms | Satellite/sensor used | Parameters for cloud detection | Accuracy |
|---|---|---|---|
| Cloud/no cloud | MODIS DATA | Discriminant analysis and support vector machine | 70%, 90% |
| Cloud/no cloud | GaoFen-1 WFV | Fusing multi-scale convolutional network | 97.85% |
| Cloud/no cloud | Spot 6-7 satellite image | ANN is applied to raw pixel values, ratio features, Gabor features, DCT features, super pixels, and patches of the images | 86% |
| Cloud/no cloud | GOSAT-2 CAI-2 | Support vector machine | 84-88% |
| Cloud/no cloud | 502 GF-1 WFV remote sensing images | Deep learning method | 85.38% |
| Thin cloud/thick cloud | GF-1 and ZY-3 images | Probabilistic latent semantic analysis and object-based machine learning | 90% |
| Cloud/no cloud | MSG-SEVIRI imager | The spectral and textural features of the MSG- SEVIRI images and physical test— estimation of class likelihoods using K-NN classifier | 90.6%, 91.2%, 90.5% |
| Thin cloud/thick | GaoFen-1 (GF-1), GaoFen-2 (GF-2), and | Multiple convolutional neural networks | 95-98% |

| Thin cloud/thick cloud | GF-1 remote-sensing images | pre-processing, seed extraction, region growing step | 93% |
|---|---|---|---|
| Thin cloud/thick cloud | MODIS DATA | Feature level fusion random forest (FLFRF), decision level fusion random forest (DLFRF) to incorporate visible, infrared (IR) and thermal spectral and textural features (FLFRF) | 82% |
| Cloud/no cloud | RASAT and Gokturk-21 | Tree structure and a multiple feature detection method using SVM classifier | 98% |
| Cloud/no cloud | Meteosat | The fuzzy logic method, the neural network method | 84.41%, 99.69% |
| Ice/snow cloud | MOD021KM-L1B data | Boosting and random forest (RF) for fusion of visible—infrared (VIR) and thermal classifiers | 98% |
| Ice/snow cloud | Sentinel-2A | Object-based convolutional neural networks | 92% |
| Cloud/no cloud | Proba-V instrument | A supervised pixel-based classification | 93% |

The methods used in the classical algorithm approach must follow known basic steps for a specific input image, while the methods used in the machine learning approach use AI to help the systems learn automatically on the basis of existing data. Both have different methods and each method has its own advantages and its own merits. There is a possibility of a false output in a classical algorithm if a wrong move is chosen. In the machine learning method, this can be contradictory due to the reliance on input data. In order to improve the precision and accuracy of cloud detection methods, the use of the machine learning approach and the classical algorithm approach will be recommended in combination as one tool.

## 4.7 New Cutting-edge Technologies

The use of satellite technology will provide an invaluable means to detect and monitor plastic in the ocean. However, as mentioned in section 2.4, some limitations make satellite technology alone insufficient to accurately track plastic debris world-wide. The resolution available with most operational satellites is such that small pieces of plastic and microplastic remain undetectable from orbit (Watanabe, Shao and Miura, 2019). Therefore, while satellite technology provides an important contribution in the fight against plastic pollution, it will have to be combined with land or air borne technologies for a complete and reliable plastic monitoring scheme.

In this section, technologies that can complement the limitations of satellites will be explored. In particular, underwater plastic, pollution in regions covered by clouds and microplastics are of interest. The objective of this search is to find suitable candidate technologies that can be integrated in a large scale solution to monitor plastic in the ocean. It is believed that a combination of satellite and non-satellite technologies, able to communicate and operate as an integrated system, will provide the most reliable solution.

### 4.7.1 Underwater Robots That Need to Communicate With Satcom & Satnav

This section will focus on those technologies that operate underwater, such as submarine drones and robots, that can be used to inspect and eventually remove, plastic under the surface of the ocean or under thin ice. These are of interest because they can be used for in-situ monitoring of plastic in the ocean, complementing EO technology. These submarine robots, often referred to as AUVs, could provide a means to monitor plastic in those places where satellite technology lacks efficacy. This includes small (potentially micro-) plastic and trash under the surface of the water or under ice.

The development of these vehicles is a result of a large interest in exploring the underwater world by scientists, environmentalists and academics. Examples of underwater robots that have been proposed, built and/or used to explore the ocean include ISE Explorer AUV, pictures in Figure 47a. This AUV is capable of diving as deep as 6000 m and can survey an area in the ocean for up to twenty-four hours on a single charge (Navingo, 2018). Explorer has been used to explore the waters under the Arctic Ice in 2010 when it operated autonomously for more than 10 days collecting data on the ocean seafloor (Kaminski, et al, 2010). Figure 47b shows a different robot, also able to explore the seafloor but based on an intrinsically different technology. This robotic crab, called *Silver 2,* was developed at the BioRobotics Institute at Scuola Superiore Sant'Anna. This AUV was built with the intention of collecting debris on the seafloor, a problem not tackled by large organizations such as The Ocean Cleanup, which focuses primarily on floating plastic (Abrams, 2019).

(a) *ISE Explorer AUV is an underwater robot manufactured by ISE, Canada.*     (b) *Silver 2: a robotic crab designed to collect plastic on the seafloor*

**Figure 47**: *Examples of AUVs that have been used in open waters. The two examples show different technologies based on separate principles, depending on their respective application (Abrams, 2019).*

These are only a few examples of the many AUVs under development across the field. While many of these technologies could not be directly applicable to the monitoring or clean-up of plastic in the ocean, they embody technological solutions that could be used for such endeavours. An AUV swarm capable of functioning autonomously, navigating waters in a specified area in search of plastic for extended periods of time could help map plastic presence in the ocean if paired with satellites. Eventually the same combination of technologies could be used to collect such waste.

In order to understand how these vehicles, together with satellites, can provide a concrete and efficient solution for monitoring the plastic content of the oceans, clarification of the underlying technologies is required. AUVs for ocean plastic monitoring are self-powered, autonomous, underwater vehicles controlled internally or remotely that employ object detection algorithms to detect and locate plastics. They autonomously differentiate between natural features of the ocean, such as fish and algae, and plastic objects. These vehicles have been shown to have such capabilities, although some problems remain. The marine environment is extremely diverse -

light conditions can vary greatly between different locations - and the variety of objects contained in the water requires very large datasets for object recognition. The cloudiness of the water can limit camera visibility and render objects difficult to differentiate (Figure 48). Furthermore, plastic debris in the ocean are often degraded by the water and the salt, which alter their appearances. This increases the wealth of the databases required for successful object recognition.

Large efforts have been put into producing functional underwater object recognition networks. As a consequence, several openly available databases of underwater plastic images have been created. From these, numerous freely-accessible object recognition algorithms have been published. Some examples are YOLO, YOLO2, Tiny-YOLO, Faster RCNN with Inception v2 and Single Shot MultiBox Detector (SSD) with MobileNet v2 (Fulton, et al, 2018). Numerous studies have been conducted to compare publicly available object recognition algorithms for various applications. Fulton et al. (2018) trained four of the most popular object detection networks to differentiate plastic from bio content in a series of images and concluded that real-time detection of plastic using deep learning models is plausible. Figure 49 shows how such algorithms can separate biological organisms from plastic content in the ocean. If equipped with this technology, an AUV could be capable of accurately mapping plastic content in a given geographical location.

(a) *In certain conditions, visibility is relatively good and plastic is easily differentiated from other underwater features*

(b) *In other conditions, cloudiness in the water renders the autonomous detection of plastic extremely challenging*

**Figure 48:** *Examples of plastic pollution in different marine environments, showing how cloudiness in the water can create challenges for robotic object detection (Fulton, et al, 2018).*

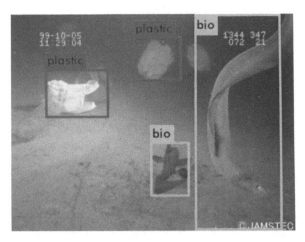

**Figure 49:** *Detection results for an image from test data for the YOLOv2 networks (Fulton, et al, 2018).*

### A Proposed Scenario of Utilization:

A scenario is proposed that could describe one of the possible applications of AUVs in combination with satellites. This scenario will highlight the potential of the two technologies when integrated and outlines a possible approach for their use in years to come.

A mothership equipped with communications capabilities with ground support through satellites carries several AUVs that have been trained using object recognition networks. The International Submarine Engineering (ISE) Explorer AUV could be a promising candidate, for example. In the scenario, these vehicles can differentiate plastic from bio presence in the ocean and they store information on the plastic amounts and on their location. To navigate in the ocean, the AUVs must know their relative position with respect to their mother ship and their relative position to each other. For this, they are equipped with basic GPS capabilities. Satellite signals can be received in the middle of the ocean, so this is a viable possibility. Underwater navigation cannot be achieved with GNSS but does not pose a technological limitation due to the extensive use of submarines. GNSS can be used to calibrate an Inertial Momentary Unit (IMU) in the AUVs which will only have to be recalibrated after the error has reached a specified value by the resurfacing of the underwater vehicles. Therefore, the AUV group would calibrate their IMU, submerge themselves to inspect a certain area, locate and map debris and periodically resurface to recalibrate the IMU.

The AUVs will periodically return to the mothership to recharge their batteries (if this is their means of producing power), recalibrate their IMUs and upload the data they have found. Satellites can then be used to relay information back to land directly from the mothership, without having to wait for the AUVs to return. This will be of great use as it will allow scientists to use data found in near real-time to direct the mothership and the AUVs where necessary based on their findings. Relaying information through satellites will also allow longer operation times as the memory of these AUVs will be limited. Given the power demand of transmitting such potentially large datasets, transmitting these from the mothership seems most suitable. This scenario would allow an autonomous operation of the AUV herd that is easily scalable (Figure 50). As the technology is developed, the same mission configuration can be applied to a herd of AUVs that can collect plastic and load it onto the mothership from transport back to land.

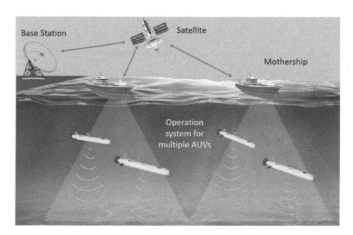

**Figure 50:** *Coordinated AUV fleet. Adapted from (Japan Agency for Marine-Earth Science and Technology (JAMSTEC), 2021).*

### 4.7.2 Zero 2 Infinity or World View Balloons

High altitude balloons are a very interesting opportunity to monitor and track plastic in the oceans, they are a way to get images of an area rapidly and possibly at a fraction of the cost compared to other techniques such as regular satellites. The balloons could also be used to drive and coordinate missions to clean the oceans.

This section discusses the products of two companies and their potential uses for this project. The two companies are Zero 2 Infinity and World View, which both produce high altitude balloons but for different purposes. A thorough assessment will determine if these technologies are suitable for the proposed solution.

**Zero 2 Infinity:**

The Zero 2 Infinity company is developing three different concepts of stratospheric balloons for different purposes, those concepts are Elevate, Bloon, and Bloostar (Zero 2 Infinity S.L, 2016) and are elaborated further in this section.

**Figure 51:** *Elevate concept being tested (Zero 2 Infinity S.L, 2021). Test payload being sent to near space for a client.*

The Elevate concept can be seen during testing in Figure 51, it is a stratospheric balloon made to go up forty km of the altitude above ninety-nine per cent of the mass of the atmosphere with a payload up to 6000 kg such as, experiments, telescopes, antennas, electronics, products and cameras (Zero 2 Infinity S.L, 2016). The mission can last from a few minutes with just an elevation and a descent to a few days depending on what is needed by the client, when it is time to come back down, the payload is released and makes a controlled landing using a parafoil (Zero 2 Infinity S.L, 2016).

The Bloon (Figure 52) concept is a stratospheric balloon that uses roughly the same technology and design as the Elevate, it is supposed to take a pressurised cabin for a crew of six and possibly measurement instruments, it is planned to stay in the air for 4.5 hours in total, reaching thirty-six km in altitude and coming back the same way as the Elevate (Zero 2 Infinity S.L, 2016).

**Figure 52:** *Bloon concept (Zero 2 Infinity S.L, 2021).*

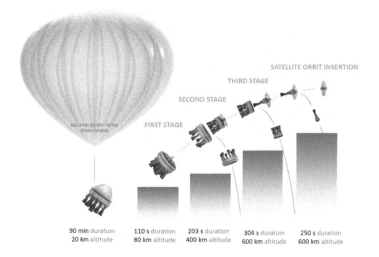

**Figure 53:** *The Bloostar concept (ESA, 2018).*

The Bloostar is a concept for an innovative launcher that uses a balloon to lift a rocket directly to the stratosphere (Figure 53), to 20-25 km of altitude where the launcher, optimised for the near-vacuum at that altitude, will ignite to send a payload into space with minimal drag (Zero 2 Infinity S.L, 2016; ESA, 2018). The advantage is that it allows the payload to have a larger volume and a different form factor as there are fewer constraints due to the absence of drag and vibrations.

Some of these technologies have been seldomly tested and are not necessarily the most reliable due to a lack of data. The price for all those technologies is quite low and are also ecologically friendly (nearly no fuel consumption compared to rockets), noise-free, and safe since the balloon is in helium and will not explode (Dassault Systèmes, 2021).

The Elevate could be used to monitor the oceans with any type of payload and also relay data to the ground and to drones on the surface of the water. Drones could further be coordinated by the issuance of manual navigation data to direct them toward nearby garbage patches. The Bloon concept is a crewed Elevate and would not bring any significant new capabilities and will not be considered since sending humans in the stratosphere is beyond the scope of this project. The Bloostar concept could help send constellations of smallsats in high inclination orbit in order to have high visibility of anywhere on the globe including the oceans but for a low launch cost.

**World View Balloons:**

World View builds high altitude uncrewed balloons composed of a balloon, a payload, and navigation and power systems such as the one that can be seen in Figure 54. Its aim is to do observations from the stratosphere as it is able to give a continuous bird's-eye view of the area under it from between fifteen and twenty-four km of altitude with a five cm resolution in real-time with low latency for up to thirty-two days according to tests (Foust, 2020; World View, 2021a).

Currently, several different payloads are available, such as imagers observing the Earth in the microwave, infrared and visible spectra. However, it is possible to design a custom payload for this balloon, allowing the customers to also see the data in real-time using a web interface after reception and pre-processing by the company (World View, 2021a). This balloon manages to approximately stay in the same area by using high altitude winds to navigate that it uses at its advantage by changing its elevation by playing on the inflation of the balloon (World View, 2021a).

**Figure 54:** *World view stratollite (Foust, 2020).*

This balloon can therefore be used for mostly stationary continuous observation of the same area which could be useful to monitor plastic in a certain area for a few days at a time or for coordinating cleaning efforts with robots. A good thing about it is that one can launch it from basically anywhere on Earth or in the oceans without a heavy infrastructure to support it which is something desirable when monitoring the oceans. Each launch would cost around $20,000 (Wall, 2015), which is extremely cheap compared to other high altitude means of observation for that long in the air. However, it must be taken into account that the cost of a launch from Arizona is not the same as one near an ocean. Both launch facilities and recovery costs need to be taken into account. Since the launchpad can just be the bridge of a boat, this problem can be solved relatively easily, keeping in mind that this will only cost a fraction of the cost of a satellite launch and have a smaller footprint on the environment.

There are a few technical limitations with that technology, indeed the area swept by the balloon is limited by the altitude and the slow nature of the balloon and that also varies depending on the winds as seen in the next figure. Few tests have been done and during some of them, the balloon ruptured causing a failure of the testing mission.

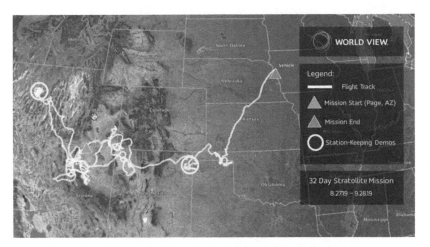

**Figure 55:** *This diagram shows the flight path of World View's Stratollite high-altitude balloon system during a test flight, which took place from Aug. 27 to Sept. 28, 2019 (Wall, 2015).*

**Discussion of the Use of Balloons to Monitor the Oceans:**

The idea of using stratospheric balloons seems reasonable to complement satellite observations for the moment when one needs continuous observation over one area which can be more useful than satellite observation due to the number of satellites required to have a similar frequency, this is minimised in the polar regions where a satellite in a polar orbit will have a high frequency of revisit which is good for EO.

The stratollites could be useful to coordinate autonomous drone missions locally. Since one can send almost any type of payload, one can combine instruments for EO and communication to control a swarm of robots. Using the stratollites as a local positioning system and measuring the relative position of the drones compared to the microplastic debris means that it is possible to optimise the cleaning of a large area with the data from the balloon even if the GPS cover is not the best in certain areas such as the poles.

As said previously balloons are ecological (compared to rockets), silent, cheap, and do not require a specific launchpad, this means they can be launched anywhere either in the ocean or near it, and are easy to build. However, balloons have a severe drawback; they will not stay in the air for more than a few days compared to years for satellites.

## 4.8 What About an Integrated Strategy for the Arctic Ocean?

### 4.8.1 Strategy

As highlighted in the previous sections, there are many obstacles to monitoring ocean plastic in the Arctic. Depending on our monitoring strategy we will encounter different technical obstacles, notably being the presence of ice and clouds. Whilst it is true that plastic pollution accounts for a large eco-nomic loss and this polar area is a unique ecosystem that needs to be preserved, it is difficult to find adequate funding for environmental causes. Additionally, the Arctic has a unique political framework that needs to be accounted for.

Therefore, we propose an integrated strategy to monitor plastic in the Arctic Ocean all year-round using space technology, while also keeping in mind the technical, political and socio-economical limitations. Our plan is in two parts depending on the ice conditions or seasons with the objective to monitor the ocean currents to predict the formation of gyres and use of new innovative technologies associated with existing space technologies to monitor ocean plastic that is present in the Arctic. This solution on how to monitor plastics in the Arctic during all seasons is detailed in Figure 56.

The year-round, space-based infrastructure needed to support a comprehensive and effective clean-up effort is largely already in place. Satellites in highly inclined or near-polar orbits are already providing observations and could aid in coordinating surface fleets and submersible drones, such as the Landsat and Sentinel families. In the near future, satellites from both government agencies and private companies can be expected to have the capabilities required for preventative and remedial plastic pollution mitigation.

During the summer and spring season, when most of the Arctic ice is melted, is the best time to use remote sensing technology with satellites and stratospheric balloons with a RS payload. Sufficiently precise RS will allow for the creation of maps that can not only determine the sources and flow of debris but also direct cleanup efforts. The balloons will also be useful to coordinate autonomous drone clean-up missions. To monitor ocean currents the sea surface temperature will be measured using satellite sensors such as Landsat 8 (NASA) or Sentinel-3 (ESA). From these measurements it is possible to predict ocean currents as the different types of ocean circulation are interlinked.

In the winter season the ice will be an obstacle to satellite imaging due to its thickness and albedo. As such the best plan would be to use underwater robots such as UAVs or jellyfish-like underwater robots such as those developed by the engineers from Virginia Tech or the hydrogen-powered bot created by a team at the University of Texas. Satellite navigation will be a necessity for robots operating independently of a fixed base of operations. Robotics researcher Erik Engeberg of Florida Atlantic University, states: 'It is important to track their locations so that they can be retrieved after a mission' (BBC, 2019), so that the plastic collected and the machine itself does not become debris (Tadesse, et al, 2012).

Autonomous capabilities will be crucial for a large-scale and long-term solution. Plastic is not only found on the surface of oceans but are increasingly found submerged as they are broken down or biofouled (Law, S. Morét-Ferguson, et al, 2010; Simon, 2019). Surface and submersible drones are a potential solution for gathering plastics at both the water's surface and at depth. There are a number of drawbacks and limitations that must be addressed before these techniques can be employed.

As described previously, machine learning techniques will be the most vital element in effective autonomous plastic debris collection. The automated systems utilised will need to be capable of independently discriminating between anthropogenic debris and 'naturally' occurring phenomena, such as corals, lithology, wildlife and the artifacts of other conservation efforts. Machine learning holds unique vitality to the enactment of environmental protection, especially in the inhospitable Arctic environment. Moore's Law, an anecdote that has been accepted as 'scientific law' by computer experts for several generations, suggests that computer technology doubles in capability every two years, while also halving costs in the same timeframe. The consequence of Moore's Law is that computer-based technology grows in application and capacity while becoming increasingly available to end-users who lack the financial resources of a superpower-state.

In the scope of marine plastic debris clean-up, Moore's Law would suggest that the computational power needed for accurate discrimination between plastic debris and non-plastic objects will be soon available to the smaller-scale budgets that environmental projects are often resigned to. Such capabilities should therefore become highly attainable by the time this proposed project is implemented, enabling the autonomous segment that is expected to operate below the surface.

To avoid interfering with wildlife, effective remediation efforts should put special effort into not harming natural life, especially as Arctic waters become harbors for increasingly endangered species. Cleanup techniques must prohibit the ingestion of litter by wildlife and avoid causing further harm to the environment. Concentrating plastics at a single location may present an increased risk of ingestion for at-risk organisms, particularly for species that display strong preferences for certain colors and types of plastics (Gove, et al, 2019). Fragmentation over time is another concern that should be mitigated for any long-term solution: the longer the plastic debris remains in the water the more it fragments into smaller and less manageable pieces. Prompt collection methods will have to be enacted to reduce the peril of propagating microplastic pollution.

The BioRobotics Institute's Silver-2 crab bot is only expected to operate at speeds of two km per hour or less (Abrams, 2019). This is a generous top speed for a legged automaton working underneath the ocean. Comparable legged robots can traverse smooth terrain at 3-4 km per hour (Dupeyroux, Viollet and Serres, 2019). The Silver-2 prototype currently has a battery lifetime of seventeen hours. Despite modern robotics technology advancing in leaps and bounds, higher locomotive speeds and battery capacity cannot be assumed to advance greatly before an Arctic Ocean cleanup begins. This severe limitation to the operational range will significantly reduce the effectiveness of the proposed autonomous cleanup efforts.

Solar power is the ideal option to keep the machines powered in the remote Arctic environment, however this will create an additional weight burden for the machines and would not be effective during winter when there is little to no sunlight at all in the Arctic Circle.

Autonomous navigation schemes will be another hurdle for robotic clean-up efforts. While GNSS-dependent techniques are effective for most areas of the world, the Arctic Circle is another feat entirely. The high latitude, deep seawater and thick ice coverage inhibit effective communication with most navigational satellites. Alternative options to augment or substitute for GNSS include Inertial Navigation Systems (INS) and Visual Navigation Systems (VNS).

Inertial navigation is in principle a measure of how far a machine has traveled in a certain direction over a given time. Also sometimes referred to as 'dead reckoning,' INS are highly effective for short range navigation. However, they quickly lose effectiveness over time, especially with legged robots. Such navigation methods are common for modern autonomous vehicles. However, any small navigational error quickly turns into a very large navigational error (Zhang, Xu and Wu, 2016) without correction provided by alternative navigational techniques.

VNS relies upon the machine's ability to recognise its surroundings to establish a position. The practicality of such a scheme is only valid over a short period of time in an environment as volatile as the Arctic Ocean. Glacial tilling, or the movement of rocks and sediment due to the movement of ice, holds tremendous sway over the Arctic terrain. Every year, every season, the landscape changes due to the wind, waves and ice dynamics that dominate the environment of the Arctic Circle. These erosional forces will drastically change the navigational environment of the Arctic Ocean and must be anticipated for every season.

Corrections available to these navigational schemes include the use of celestial compasses. A nonmagnetic compass that determines geographic North from the ionization of ultraviolet radiation in the Earth's atmosphere is not only reliable and allows for navigation seventy times more accurate than civilian GPS technology alone but remains accurate in all weather and even when used under water (*loc. cit.*). This emerging, biology-inspired technology is worth considering for any autonomous vehicle being proposed for use in the Arctic Ocean (Dupeyroux, Viollet and Serres, 2019).

Active surface cleaning techniques such as the Fishing for Litter Fleet (KIMO International, 2019a), the Floating Robot for Eliminating Debris (FRED) and the OCP (Clear Blue Sea, 2020) are capable of cleaning tremendous amounts of debris, despite still largely being underfunded or in the prototyping stages. The best strategy to deal with the pollution problem is one that combines the active and passive, autonomous and crewed methods for a comprehensive and aggressive solution. With the Arctic Ocean being estimated at having the lowest volume of debris of the world's oceans, it makes the ideal environment to test a combined solution to the issue of plastic pollution.

Assuming that the autonomous members of our integrated strategy will need to be cleaned and maintenanced regularly, it would probably be most feasible to have them and their support crews operate primarily in the summer and fall months when ice is at a minimum. This reduces the operational risks for both crewed and autonomous methods. Navigating the ocean will be far safer and easier when sea ice is at a minimum and visibility is high. Sea ice is also likely to trap debris during winter, reducing the effectiveness of collection efforts (Kylin, 2020). When ice retreats in the spring it releases the trapped debris and carries more debris from the land to the ocean (*loc. cit.*). Cleanup efforts will need to anticipate these seasonal changes.

This integrated strategy will need to operate in conjunction with policy and law changes that protect the Arctic and endeavor to reduce plastic waste from continuing to freely flow into the oceans. The lower volume of ship traffic in the Arctic Ocean should ease the enforcement of existing laws, such as those prohibiting the dumping of plastic at sea (IMO, 1988). The monitoring of GNSS transponders and satellite observations permit for the monitoring of illegal dumping and fishing without requiring 'ground' observers (Tan, 2020). Similarly, the shipping and fishing fleets that seek to operate in the Arctic could be required to maintain GNSS transponders on their large nets and cargo containers, providing information on losses and a means to track large and potentially valuable or harmful debris (Peeters, 2021).

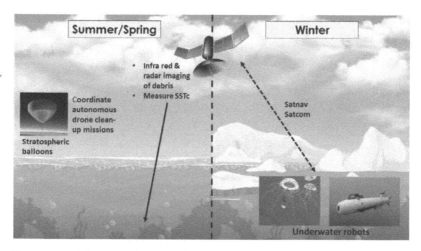

**Figure 56:** *Schematic diagram of the strategic integration of various space technologies with new innovative technologies to track, monitor and coordinate clean-up of plastic debris in the Arctic all year round.*

## Cost Estimate of the Operation

From this a preliminary estimation the cost for the monitoring operations of plastic pollution in the Arctic for a variety of solutions can be made. The current cost of using a Zero 2 Infinity balloon is $132,000 per tourist (Zero 2 Infinity S.L, 2020) - it is in theory possible to make a commercial deal with the company in which the enacting of the solution would only require paying to attach a payload to the balloon.

The total cost will depend on the instrumentation used but these could cost anywhere from $5,000 to $100,000 each in addition to the daily operation cost of approximately $880 per day (Larson, 2014). The advantage of using an AUV is that they can be repurposed for other uses as they are very versatile, upgradeable and reconfigurable (*loc. cit.*). The cost of AUVs on the market are approximately $10,000 for purchasing but some low-cost models are being developed with a price tag at $400 (Fowler, et al, 2016). The difficulty will be having a network that can provide communication with the end-user and between the AUV unit in operation.

**Table 8:** *Existing and near-future satellite sensor systems of relevance for inland water quality in Australia. Adapted from (Dekker and Hestir, 2012).*

| SATELLITE SENSOR SYSTEMS | | PIXEL SIZE (M) | SPECTRAL BANDS (400–1000NM) | REVISIT CYCLE | RAW DATA COST PER KM2 (AUD)[a] | WATER QUALITY VARIABLES [b,c] | | | | | |
|---|---|---|---|---|---|---|---|---|---|---|---|
| | | | | | | CHL | CYP | TSM | CDOM | K[e] | TURB SD |
| *Current Ocean-Coastal Low Spatial Resolution* | MODIS | 1000 | 9 | Daily | Free | ● | ● | ● | ● | ● | ● |
| | MODIS | 500 | 2 | Daily | Free | ● | ● | ● | ● | ● | ● |
| | MODIS | 250 | 2 | Daily | Free | ● | ● | ● | ● | ● | ● |
| | MERIS& OCM2 | 300 | 15 | 2-3 days | Free | ● | ● | ● | ● | ● | ● |
| | VIIRS& JPSS | 750 | 7 | 2x/day | | ● | ● | ● | ● | ● | ● |
| *Current Multi-Spectral Mid-Spatial Resolution* | Landsat | 30 | 4 | 16 | Free | ● | ● | ● | ● | ● | ● |
| *Current High Spatial Resolution*[d] | IKONOS, Quickbird SPOT-5, GeoEYE | 2-4 | 3-4 | On-Demand 2-60 days | 5-15 | ● | ● | ● | ● | ● | ● |
| | RapidEye | 6.5 | 5 | Daily | 1.5 | ● | ● | ● | ● | ● | ● |
| | Wordview-2 | 2 | 8 | On-Demand | 30 | ● | ● | ● | ● | ● | ● |
| *Future Ocean-Coastal Low Spatial Resolution* | Sentinel-3 | 300 | 21 | Daily | Free | ● | ● | ● | ● | ● | ● |
| *Future Multi-Spectral Mid-Spatial Resolution* | LDCM | 30 | 5 | 16 | Free | ● | ● | ● | ● | ● | ● |
| *Future Hyper-Spectral* | EnMap | 30 | 90 | On-Demand | Free (?) | ● | ● | ● | ● | ● | ● |
| | PRISMA | 20 | 60 | 25 days | Free (?) | ● | ● | ● | ● | ● | ● |
| | HySpIRI | 60 | 60 | 19 days | Free | ● | ● | ● | ● | ● | ● |

● Highly Suited    ◉ Suitable    ◑ Potential    ● Not Suitable

CHL=Chlorophyll; CYP=cyanobacterial pigments such as cyano-phycocyanin and cyano-phycoerythrin; TSM=total suspended matter; CDOM =coloured dissolved organic matter; $K_e$= vertical attenuation of light coefficient; TURB= turbidity; SD=Secchi Disk transparency

[a] Raw data costs are per image. Bulk acquisitions may attract a discount.

[b] Products in development are: coarse particle size distributions and phytoplankton functional types.

[c] Model-management integrated products under research are: eutrophication index, water quality index, algal bloom index, carbon content/flux and contaminant estimation.

As seen above in the table from Dekker and Hestir (2012), there are a variety of satellites available to begin EO for ocean protection purposes. The United States Geological Survey (USGS) LandSat-8 appears to be the best option for detailed observation and the ESA Sentinel-3 best suited for comprehensive oceanographic data collection. Other options that could be utilised (for a cost) include IKONOS and GeoEye. In summary: LandSat series of satellites is the best option to begin ocean cleanup efforts for low cost. The fast-growing field of EO will present newer options with superior resolution.

### 4.8.2 Scalability to Other Oceans

The technologies selected for the plastic monitoring strategy of the Arctic in section 4.8.1 were also chosen based on their use in waters besides the Arctic ocean; therefore, this monitoring strategy can be scaled to other oceans with relative ease. In fact, monitoring the Arctic ocean presents more challenges than the rest of the oceans on Earth due to the location's geophysical characteristics. For these reasons, the proposed approach applied to other oceans will be simpler, less costly and more effective.

The absence of ice in oceans such as the Pacific or the Atlantic makes the communication between the AUVs and the satellites easier. In the Arctic Ocean, AUVs had to resurface to relay data and to recalibrate their IMUs. This will no longer be necessary in the absence of ice, which reduces the complexity and the price of these robots. Furthermore, the illumination in any other ocean is better than in the polar ones, which allows for better reflectance reading. Most satellites are in near-equatorial orbits so will have better coverage and better temporal resolution of other oceans than the Arctic Ocean (e.g. Sentinel-2 satellites). A lower cloud coverage over the other oceans will allow passive detector satellites to easily monitor and track plastic in the ocean. Stratospheric balloons can be deployed from land in Europe or in the United States directly, as opposed to being deployed from ships in the open sea. Atlantic and Pacific oceans are extensively studied, so more data will be readily available (regarding currents for example). This will result in a more targeted deployment of AUVs. Moreover, the water temperature in those areas will be milder and therefore AUVs batteries will last for a longer period of time for each deployment. For all these reasons, it is believed that the strategy of monitoring and tracking plastic in other oceans would be cheaper and more effective than for the polar oceans.

## 4.9 Chapter Summary

Chapter 4 offered a handful of monitoring and tracking methods focused on the quantitative and qualitative analysis of chapter 2. These analyses concluded that a space EO method that uses FTIS is the optimal solution for the detection of aggregated floating plastic debris in oceans. This method is possible due to the chemical properties of plastic. It absorbs or emits light under complex infrared radiation conditions that are different from the spectral bandwidth of the water. SAR provides a fast way to search oceans for large floating debris patches because it has a low spatial resolution. In addition, EO satellites are able to provide a reasonable coverage of up to 90°N when a sufficient orbit and spatial resolution are provided. In addition, data can be retrieved free of charge from certain satellite providers, making the solution even more appealing.

As can be seen from EO's specifications and constraints on satellites, a single space solution is unlikely worldwide. Indeed, as the example of the Arctic Ocean shows, the seasonal cloud and ice coverage restrict or make EO impracticable from space. Fortunately, some technologies could complement the EO very well. Therefore, the section took a closer look at the AUVs and the high-altitude balloons. The integration of these complementary technologies and the EO satellite is provided in the form of an integrated solution. A fundamental cost analysis of this idea is also presented and shows that the integrated model is achievable.

# 5

# And Action:
# Community Education And Investment

---

*'Our actions over the next 10 years will determine the state of the ocean for the next 10,000 years.'* - Sylvia Earle, *marine biologist, oceanographer.*

## 5.1 Introduction

In 3.3, scientific diplomacy was determined as a means of facilitating cooperation among states, particularly through the Arctic Council. Science has the power to investigate phenomena that adversely affect the entire ecosystem and has the possibility to increase awareness amongst different stakeholders. Not only can it educate and influence Arctic State members but it can reach a wider audience including younger generations, educators, scientists, and corporations. Nonetheless, science risks remaining self-referential if not supported by a means of communication that makes it accessible to a greater number of people. That is why it is important to identify first of all the different types of audiences to which plastic pollution awareness would most influence before identifying the most appropriate means of communication.

In this chapter, various categories including; government or policy-makers and the general public - including young generations - are investigated, in order to design outreach tools that may be useful to raise their awareness regarding plastic pollution in the ocean. The aim is to make them understand two separate but complementary messages for a common purpose: a major reduction in marine plastic pollution.

The first audience targeted are the policy-makers, who ultimately decide whether to invest financial and human resources to implement a political and economic strategy that tackles the problem. The second audience targeted is the general public, to communicate the importance that everyone can make an effort to follow best practices and contribute to that strategy. Finally, an initiative championed by the International Space University (ISU) is proposed involving both alumni and faculty to physically take action against plastic pollution.

## 5.2  A Virtual Symposium on Marine Plastic Pollution

As mentioned, there are two ways for providing outreach and communication about the current situation of plastic pollution in oceans. The first one is dynamised by governments (Figure 57) to exercise their role as decision-makers. Governments can actively influence industry behavior by prohibiting the dumping of their wastes in rivers, or by encouraging the production of new biodegradable materials. Such an awareness campaign could lead to numerous trade-offs in many different sectors, including the spontaneous birth of new eco-sustainable initiatives within various companies, from the largest to the smallest.

Using their decision-making power, policymakers may also enhance science classes based on ecoresponsibility, from primary school to high school, supported by the intervention of external experts, such as scientists and activists. The aim is to raise awareness about the pollution in oceans among new generations, to encourage them to be more respectful of the environment and become more aware of the effects of their consumption. Governments also have the power to organise the dissemination of specific news stories through the mass media, with the aim of influencing the general public who, in this way, may become more aware of them.

In order for governments to positively influence other audiences on the need to solve the problem, they themselves must first be influenced. An effective way to engage governments is to inform them of the extent of the problem and the various existing opportunities aimed at solving it. For example, interest groups could organise an international virtual event on the specific topic of plastic pollution in the ocean and invite government officials. The idea is to design a virtual fair along the lines of the International Astronautical Congress (IAC) 2020, providing a range of activities and interfaces with the ability to attend both live and pre-recorded events.

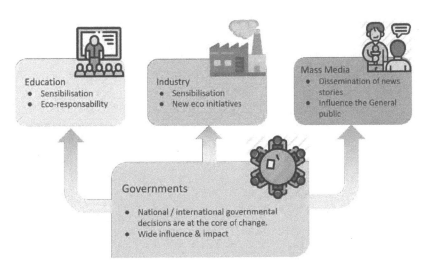

**Figure 57:** *Governmental communication and outreach scheme.*

The government of Iceland, for example, has organised the International Symposium on Plastics in the The government of Iceland, for example, has organised the International Symposium on Plastics in the Arctic, an online event in collaboration with multiple partners such as the UNEP, in order to inform governments about the threat of plastic to ocean life. Our idea would be to create a similar event, using the format of a virtual fair, as IAC2020 did, dedicated exclusively to governments and policymakers, but with the participation of scientists, industrials, start ups and other professionals as panelists. These experts would bring their experience as an example of best practice, possibly replicable in other countries.

The event could be spread over three days, with the first day dedicated to assessing the problem, the second to monitoring and the third to mitigation measures. Each day, therefore, the same virtual hall would host different sessions, courses, and workshops, depending on the focus, with a central area, always open, and dedicated to discussion among participants. This would foster a creative environment aimed at engaging participants in discussing solutions or even creating partnerships with each other. Below is a first schematic representation of what the virtual room might look like, including some ideas about the sessions for each day.

The key message that the event would have to get across to decision-makers is the importance of investing human capital and financial resources in a long-term strategy, including monitoring and mitigation techniques, to solve the problem of plastic pollution in the oceans.

In order for such a symposium to gain traction, it may need to be organised or sponsored under the auspices of a UN organization relevant to the issue, such as the International Maritime Organization (IMO). The event could also be sponsored by other international governmental and non-governmental institutions, with participation from the private sector, including some start-ups, that could share their expertise with government representatives. One outcome could be a commitment to creating a working group on marine plastic pollution at the United Nations General Assembly, which would work to explore international solutions to the problem since plastic pollution can begin in one location and end up on the other side of the world.

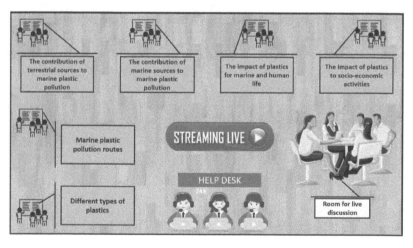

**Figure 58:** *Day #1: Assessing the problem of Plastic Pollution in the Oceans.*

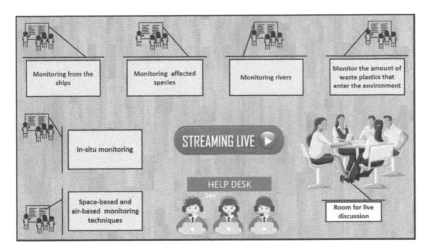

**Figure 59:** *Day #2: Monitoring Plastic Pollution in the Oceans.*

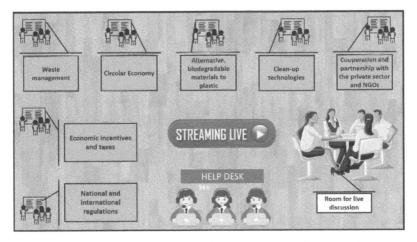

**Figure 60:** *Day #3: Mitigating the problem of Plastic Pollution in the Oceans.*

## 5.3 Social Media

Alongside governments, the other major driving force capable of influencing concrete action to solve the problem is the general public, including younger generations. In this case, communication can come from an individual (e.g. Greta Thunberg), an informal group, an expert group, an NGO, or through peaceful movements and demonstrations (e.g. Extinction Rebellion). Through social networks, anyone can be a journalist, showing the extent and risk of plastic pollution in the oceans, raising awareness and up to prompting government action.

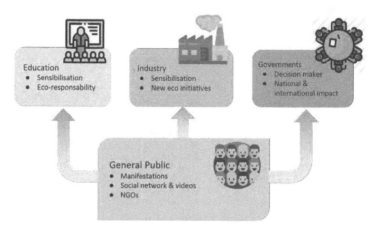

**Figure 61:** *General public communication and outreach schema.*

However, as with governments, it is, first of all, essential to draw the attention of masses and the youth towards the rising concern of marine plastic debris in the oceans. It must be realised that human activities not only harm the marine environment but pose a threat to humankind in the present and long run. If plastic pollutants persist, given the current environmental crisis, it won't be long until our home planet is destroyed. This is the message that complementary to the previous one, should arrive to the general public and to the young generations.

There are several projects that are already implemented within primary and secondary schools to raise awareness of the problem. One example is the 'fish or plastic ?' project, carried out by students of the Luxembourg High School 'Lycée Athénée du Luxembourg'. They created a series of aquarelle paintings. Through drawings (shown in Figure 62), the students have tried to raise awareness of the marine plastic pollution problem by exhibiting them at L'Ascenseur Elevator Plateau St. Esprit Grund, a well-known tourist location in Luxembourg.

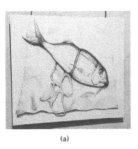  

(a)                              (b)                              (c)

**Figure 62:** *Drawings from the 'fish or plastic?' project, made by the students of the Luxembourg High School 'Athénée' (Antoni, et al, 2020)*

Social media platforms such as Instagram, Facebook, and Twitter act as the most influential systems of articulation. Having said that, we propose a set of trending hashtags that could be accompanied by posts or initiatives with context to plastic pollution in order to resonate with the youth. The hashtags #space&oceans #sustainableoceans&me #NoPlanetB #TeamOceans are some official ones from our team. These hashtags could initiate just another trend among the youth and even a mere step towards mitigating plastic pollution will make a difference on a global level.

## 5.4  SpaceCleaning. Get Rid of Plastic Pollution with ISU!

As MSS 2021 students of ISU who are much aware of the rising concern of plastic debris, we would like to initiate an annual ISU Plastic Cleanup event that will include both students and faculty to be proactive towards the environment and become aware of ISU's contribution to plastic pollution. These cleanups can be along the Rhine river, given that it can eventually end up in the North Sea. The North Sea connects to the Arctic Ocean, therefore literal upstream cleanup efforts would help prevent downstream Arctic plastic pollution. Thus far, ISU will be humbly contributing to the mitigation of plastic pollutants around its surroundings as well as the Arctic Ocean.

The venues can vary each year depending on the plastic pollutants spread across countries and islands. The ISU plastic cleanups can be a collaborative initiative with existing NGOs like the Good Karma Projects or even other students on Innovation Campus to include more pupil on this moral cause. To kickstart this project locally, we can join hands with the Université de Strasbourg and expand this initiative eventually.

**Figure 63:** *ISU clean-up initiative logo.*

The plastics collected during the cleanups will be sent forth to recycling units or known environmental conservation organizations like the Oceanmata in Germany who share a vision to rid the oceans of plastic waste. In alignment with ISU's 3I's principles (International, Intercultural, Interdisciplinary) and collaborations, this initiative could take a turn of events and soon be a proactive face of the generation.

## 5.5 Chapter Summary

To conclude there are two main audiences focused on for the educating and disseminating information about global marine plastic pollution. The first one is by governments and policymakers themselves who can utilise their power and role as decision makers to influence industries, groups and individuals in taking plastic pollution seriously. Governments have the ability to direct the educational sector, plastic industry and media in letting a global issue be known (Figure 57).

The government needs to be advised to accurately take a decision and propose measures, therefore the idea is to propose an international symposium, based on the IAC2020 model, as it has been done by the government of Iceland which advocates the danger of plastic wastes in the Arctic. The idea would be to provide virtual events, with the participation of scientists, industrials and experts, who would bring their observations and recommendations to mitigate the problem of plastic solutions to the government worldwide. It could take place in three days: assessment of the problem, monitoring ways, and mitigation measures. One day for each, with several activities, courses and workshops according to one specific focus.

The second way to convey information is through the general public, by means of manifestation, social networks or NGOs. As explained above, the younger generations are already impacted and play an important role as a vector of sensibilization, with for example the project 'Fish or Plastic' from the Luxembourg high school, Athénée. Space and Ocean Team project as well, as students who are facing those problems want to support the awareness, across the realization of an education short movie. Inside of what, numerous plastic alternatives and mitigation techniques will be promoted. The Team could mobilise its owns hashtag #Space&Oceans and #TeamOceans, which can help make the topic a trend and inform people all over of the dangers of incorrect plastic disposal methods.

Furthermore, the Space and Ocean team would like to launch an initiative called ISU Plastic Cleanup, inviting both staff and students to organise cleanup drives, alongside beaches or the Rhine river. Collected plastic would then be sent directly to recycling centers. The use of educational tools, social media and audiovisuals can help inform people of all ages and industries on the importance of mitigating the plastics that enter the oceans because ultimately the goal is to prevent plastic from entering the oceans from the start.

# 6
# Putting It All Together:
# Conclusions and Recommendations

'The sea is the only embodiment of a supernatural and wonderful existence.' - Jules Verne, *novelist, poet, playwright.*

## 6.1 Conclusions

This report takes an interdisciplinary approach to investigate 'Tracking Plastic Pollution in Oceans from Space'. Since microplastics originate from macroplastic and because there is currently no feasible way to analyze microplastics pollutants remotely from space, it is much more practical that this report concentrates on macroplastic treatment. The report explores the subject through the foundations of four main pillars that narrows the scope of a successful LR. Following the extensive research on plastic pollution, the Arctic Ocean, space technologies and science communication, this section will draw conclusions under each of these four chapters.

### 6.1.1 Plastic Pollution in the Oceans: Problem Review

There is an enormous amount of plastic accumulating in the major five gyres across the globe with millions of tons of plastic flowing into the ocean each year. There are deleterious effects of such pollution on biological life, ranging from zooplanktonic organisms to both invertebrates and vertebrates, including humans. Biochemical pathways of plastics within such organisms are not as well characterised as they could be, but what is known is that plastic is not easily 'flushed' from the organisms. This may pose risks to organs, germane the mucus membranes and filtration structures of the body within humans, such as the lungs and kidneys. Plastic can cause acute

131

or chronic inflammation, which may contribute to other possible health problems, such as gastritis, COPD and IBD. Moreover, plastic pollution directly impacts local populations that depend on revenues received from fishing and tourist seasons since plastics impact fish livestock and fewer tourists go to polluted areas.

Technologies are vital in assisting with solving the issue and are used in two main ways, collection and prevention of plastic. Plastic pollution is tracked and monitored using non-space related methods such as modelling approaches to monitor rivers and direct and indirect tracking methods to identify pollution routes and track macroplastics (e.g, in situ plastic clustering sampling using NDVI and FDI), varying sampling locations and also modification of the FTIS. Currently, existing solutions allow for the determination of plastic patches in the oceans and rivers with a high accuracy rate. Plastics can also be monitored and tracked using space and airborne technology and have been detailed in this report such as: Reflectance spectroscopy, 'Clever-Volume' system using a LIDAR technology to assess waste management systems from space, GPS active trackers to map plastic routes and Sentinel-2 satellites are used to track and monitor plastic. AI and machine learning will play a fundamental role in assessing the plethora of data provided that will be utilised in plastic mitigation.

### 6.1.2 Plastic Pollution in the Arctic Ocean

This report provides an overview of the formation of the sixth gyre in the Arctic. The rise of plastics there was first reported in the 1970s, adding further concern to this area of the world already suffering from the phenomenon of melting ice sheets. The Earth's thermohaline circulation also transports marine litter from the world's other oceans to the North Pole. The natural processes that drive the loss of sea ice and convey pollution into the Arctic increase in intensity as they self-perpetuate their own causal conditions.

One of the effects of permanent sea ice loss is the possibility of opening new shipping lanes that will undoubtedly intensify human activity and exacerbate existing environmental problems, including plastic pollution. Shipping and fishing can cause large amounts of pollution and can be seen as a reflection of the inputs created by coastal populations. The report determined that with the inevitable creation and exploitation of the TSR and expansion of the NSR and NWP, plastic pollution originating in the Arctic region will increase as marine and land based activities expand. Despite current estimates that eighty per cent of plastic pollution originates from land based sources, in the Arctic where much of the marine activities are fishing, tourism, sport and research related, plastic pollution from marine sources may be greater. The increasing amount of plastic pollution not only places boats at risk but also other industries including natural resource extraction and seaport activities. It is precisely these activities that, if unregulated, can increase pollution amounts. Moreover, existing studies signified that economic centers of European and Asian countries are witnessing a slight shift upwards as Arctic opportunities expand, leading to increases in northern plastic sources. Governments such as Russia, China and the United States, have already expressed their intention to use trade routes immediately with little respect for the climate.

The other relevant consequence found in this report is that indigenous people living in the Arctic states are indirectly affected by the issue. Indigenous peoples in remote living conditions are heavily dependent on the Arctic marine ecosystem for food, trade and health. The indigenous people of the Arctic consume four times as much fish as the global average. Overcrowded fishing vessels will generate more plastic waste and create an environment where their choice food staples will become more unattainable by legal fishing techniques.

It was determined that instead of enacting an Arctic Treaty, which would come across as mainly symbolic and challenging to pass, the use of scientific diplomacy is the best course of action to engage countries and impacted subgroups in the Arctic's growing plastic numbers. The Arctic Council could facilitate the creation of public private partnerships and become an effective mechanism that incorporates the needs of indigenous populations, shipping companies, local businesses, Arctic researchers, and other interested parties. This would create a movement or driving force to ensure a habitable and clean Arctic environment.

### 6.1.3 Space Technology to Monitor Plastics in the Oceans

Chapter 4 presented a handful of monitoring and tracking approaches based on the quantitative and qualitative analysis from chapter 2. This chapter started by providing an overview of the requirements needed to perform EO with satellites. Two current applications of EO satellite monitoring are provided: monitoring the water temperature and the five gyres are mentioned since they are suspected to or correlate with marine plastic waste.

A new way to use EO satellite technology is to use the FTIS to detect aggregated floating plastic debris. FTIS is based on the physical property of chemical compositions to absorb or emit light in specific conditions in the field of infrared radiation. The possibility to distinguish between the different objects on the water surface based on its spectral bandwidth was analyzed by Topouzelis, Papakonstantinou and Garaba (2019) and by Biermann, et al. (2020). Another strategy is to use SAR, which provides a quick way of scanning the oceans for large floating debris patches but limited for smaller pieces since it has a low spatial resolution. However, SAR has the advantage of overcoming some cloudy conditions where FTIS fails. With the combination of FDI and SAR, it is possible to detect aggregated waste floating in the ocean in nearly every weather condition. Some satellites provide open-source data, making space-based approaches even more attractive.

These analyses conclude that a space-based approach for detecting plastic pollution in the oceans is a reasonable solution. Next, these applications are analyzed to their applicability in the Arctic Ocean as this report's case study. Ice and clouds render the employment of satellite-based methods more difficult but not impossible. Therefore, the strategy is to use complementary technologies, like automated underwater vehicles and high altitude balloons. An example of an integrated strategy that works for the Arctic Ocean and other oceans was given. The strategy is based on different technologies depending on the ice's seasonal situation in the Arctic Ocean. This way, it is possible to track and monitor plastic pollution primarily with EO satellites since the data can be retrieved everywhere on Earth. When the situation requires, the complementary technologies from section 4.4 will be available for monitoring and tracking. A fundamental cost analysis of this idea shows that the solution is achievable.

### 6.1.4 Outreach of the Plastic Pollution Problem

Chapter 5 discussed methods for outreach and communication about the plastic pollution problem in oceans. The weight lies with educating the wider audience outside of the scientific community.

The two audiences identified or targeted are policy-makers and the general public, with emphasis on the younger generations. On the governmental level of policy-makers, the author analyzes the idea of a virtual international symposium, based on the International Astronautical Congress model, to defend against plastic debris in the Arctic. Scientists, industrialists and experts would bring their observations and recommendations for mitigation solutions to worldwide governments. The general public outreach is done through manifestation, social networks or NGOs to increase general awareness.

The report concludes by providing an initiative championed by the International Space University involving both alumni and faculty. The initiative aims to take action against plastic pollution in person and mobilise students, professors and local populations to contribute to cleaning up coastlines.

**Table 9:** *Initial Aims and Objectives Contextualised Within Conclusions and Recommendations*

| N# | Description |
|----|-------------|
| 01 | Perform a wide-scope interdisciplinary review about the problem of plastic pollution in the oceans, including:<br>• Its sources and its consequences on society, economy and policy<br>• A review of space-based and air-based technologies for tracking and monitoring plastic in the oceans<br>• Examples of existing mitigation techniques |
| 02 | Analyze the current context of plastic pollution in the Arctic Ocean by:<br>• Describing the importance of preserving this region<br>• Discussing the outcomes of the sixth Arctic gyre on indigenous populations and economic activities<br>• Considering the best ways to coordinate an environmental preservation effort |
| 03 | Identify and analyze technologies with space or near-space components to detect, track, and mitigate plastic debris in the Arctic Ocean, by:<br>• Discussing the major challenges an ice-covered ocean can pose to monitoring<br>• Exploring cutting-edge technologies that can complement the limitations of satellites<br>• Developing an integrated strategy that can be scaled to other oceans |

| 04 | Propose a communication strategy that could raise awareness about plastic pollution in the oceans, through:<br>• A virtual symposium for governments<br>• Social media for the young generation<br>• A clean-up initiative organised by ISU |
|----|----|
| 05 | Recommend a series of considerations to make, along with some actions to undertake, in order to better face the problem of plastic pollution in the oceans. |

## 6.2  Recommendations

**Table 10:** *Recommendations - Based on the research, these recommendations were determined to further investigate the sixth gyre forming in the Arctic Ocean and determine the best ways to prevent its growth with space-based technologies being a vital tool.*

| Addressing plastic pollution in the arctic | |
|----|----|
| RECOMMENDATION 1 | **Understanding impacts on Human Health and Marine Life**<br>To conduct more research on health outcomes associated with plastic that accumulate in life forms at varying food chains levels to determine the medical implications of plastic inhalation and ingestion |
| RECOMMENDATION 2 | **Quantitative Analysis of Using New Arctic Trade Routes and Increased Plastic Pollution**<br>To conduct a cost benefit analysis of the trade offs between fuel, time, and greenhouse gas savings from new Arctic trade routes versus the costs from increased plastic pollution on marine life, shipping industries, and society |
| RECOMMENDATION 3 | **Fostering Partnerships to Assess Socio-economic Opportunities and Costs**<br>To create partnerships between public and private sector actors including industries and NGOs; actors who can bring their technical expertise to tackling this global challenge. |
| RECOMMENDATION 4 | **Arctic Council Initiatives**<br>The Arctic Council has been identified as the ideal mechanism and regulating body to push states to do more, therefore it is advised to foster scientific diplomacy between members and the inclusion of subgroups. |

| Further investigation of space technologies and applications | |
|---|---|
| RECOMMENDATION 5 | **El Niño and La Niña cycles**<br>A large-scale study to further investigate marine debris movement during El Niño and La Niña cycles to determine if weather patterns no only degrade plastics but also redistribute them in our oceans. |
| RECOMMENDATION 6 | **EO solutions for macroplastic detection from space**<br>To overcome difficulties in differentiating artificial objects and contrasting plastic from vegetation, it is recommended oversampling techniques, SAR, and NDVI optimization. Using NDVI analysis, it is not needed to analyze every single pixel, but can examine a photograph o a single area. |
| RECOMMENDATION 7 | **High-altitude balloons**<br>To further investigate the life expectancy of the high-altitude balloons and possible end-of-life scenarios since they can complement satellite observation. |
| RECOMMENDATION 8 | **An integrated Solution**<br>To integrate ground-based technologies with space-based ones to maximise efficiency in plastic pollution monitoring; it is therefore advised to investigate the different strategy component interdependencies and build a mission for the suggested solution. |
| Science diplomacy and outreach | |
| RECOMMENDATION 9 | **Government Level**<br>To have a neutral outreach program to face plastic pollution that can be used for every country. That way, it is less likely to provide a biasec message and more likely to be adopted. |
| RECOMMENDATION 10 | **Public Level**<br>The use of hashtag and social media can have a profound impact on disseminating information to all age groups, especially younger generations. |

This chapter provides recommendations based on the research conducted for this project to further understand the dynamics of plastic pollution and advance existing methods for monitoring and tracking plastic from space. The recommendations aim to mobilise communication and outreach to elucidate the extent of the problem and assist in increasing overall plastic pollution awareness in the Arctic Ocean. It is important that with the adoption of innovative methods that the origin of marine pollution can be identified, requiring the collection of immense amounts of data. Particularly through technological solutions that are capable of detecting large debris in the ocean, by locating marine pollution sources, strategies can be implemented to mitigate the pollution before it enters the oceans. Monitoring the severity of the evolving problem will be the basis of integrated action involving all sectors such as: political, scientific, industrial, NGOs and local communities. Recommendations are provided for understanding the impact of plastic pollution on marine and land ecosystems, human processes and industries. Finally, recommendations on policy actions and partnerships are given.

### 6.2.1 Addressing Plastics Pollution in the Arctic

**Recommendation 1: Understanding Impacts on Human Health and Marine life**
It is advised to conduct more research on biochemical pathways and health outcomes associated with plastic that accumulate in life forms at varying food chains levels to determine the medical implications of plastic inhalation and ingestion. Indigenous populations depend heavily on the health of the marine ecosystem for their livelihoods. Future studies should focus on filling current knowledge gaps and data gaps about the ingestion of plastic. Monitoring of biota, such as northern fulmars, is equally challenging. An optimal method of monitoring plastic accumulation in the Arctic Ocean is still outstanding. The progression of EO methods should be invested in for the monitoring of plastic in the Arctic.

**Recommendation 2: Quantitative Analysis of Using New Arctic Trade Routes and Increased Plastic Pollution**
Further quantitative analysis is necessary on the new trade route opportunities in terms of fuel and time savings and their relation to plastic pollution. Current data is insufficient to determine how much fuel is being saved. While estimates can be made on time savings using tools such as Google Earth, expanding these studies to include routes from other major ports (i.e. Murmansk) would be useful. Current studies only consider Rotterdam as a port destination (Humpert and

Raspotnik, 2012). Further, the correlation between shipping routes and plastics pollution can be thoroughly investigated as society gets closer to an ice-free Arctic.

An additional factor to be considered is greenhouse gas savings but may also be a cost if improper fuels are used. It is recommended that quantitative studies be used to examine the savings from the new routes versus the losses due to plastics. Whether it is from container losses, fishing nets, boat collisions, plastic from cruises or research expeditions, it would be beneficial to know the cost benefit analysis of using these routes and the losses from plastic debris to marine life, shipping industries and society.

### Recommendation 3: Fostering Partnerships to Assess Socio-economic Opportunities and Costs

The way society considers the economic impact of plastic pollution needs reconsideration with particular focus on adopting life-cycle costing models, which assess the costs of plastic from manufacture to landfill. It is recommended that partnerships are created between public and private sector actors including industries and Non-governmental Organizations (NGOs); actors who, once in synergy with each other and with governments, can bring their technical expertise to tackling this global challenge. This cooperation can replicate initiatives already existing in different countries and regions of the world, the so-called best practices, adapting them to the more challenging conditions of the Arctic Ocean.

Conditions in the Arctic are unique, requiring input from the different populations living there and involving them in the decision making process would first make them an integral part of the projects to mitigate the threats of ocean plastic in the ocean, and second, ensure that the mechanisms put in place to monitor and manage plastic pollution can be maintained by the locals. The need to use simple, straightforward language cannot be understated in making the difference in communication and ensuring that transparency is maintained.

### Recommendation 4: Arctic Council Initiatives

Fundamental policies are useful in raising awareness of the growing number of plastic in the ocean, but voluntary plastic policies and activities such as beach cleanups are not long-term solutions.

Governments should use regulatory, economic, and informational instruments to implement measures against plastic pollution at national level. Prioritizing scientific diplomacy and knowledge sharing is preferable to developing binding policies that would be unlikely to have compliance.

The Arctic's evolving situation requires intergovernmental regulation that only a joint effort based on cooperation can achieve; therefore, the Arctic Council has been identified as the ideal mechanism and regulating body to push states to do more, using in particular scientific diplomacy and the inclusion of subgroups. Although an Arctic Treaty is desirable, it remains more likely to work through a soft power mechanism, supported by global and national legislation. Disputes already exist over the legally binding authority of an Arctic Treaty. It is recommended that more economic and human capital resources be given to the Arctic Council, particularly for the purpose of implementing a task force to address the problem of plastics in the ocean without relying on voluntary funds from participating states. Moreover, the Amarok monitoring tool should be strengthened to monitor Arctic Council projects over the long term and evaluate their results to establish medium-term targets that allow adjustments of the current solutions are necessary.

### 6.2.2  Further Investigation of Space Technologies and Applications

**Recommendation 5: The El Niño and La Niña cycles**
Section 4.3.1 presents how El Nino and La Niña cycles impact ocean current tracking from space. These cycles cause periodic changes in the water temperature of oceans, significantly influencing the sea surface temperature, which can be observed via satellites. Although little evidence was found, the cycles could be contributing to the movement of marine debris in the Pacific and elsewhere. Since only limited information about the alleged relationship between ocean temperature and plastic pollution is currently available, it is recommended that a large scale study take place to further investigate marine debris movement during El Nino and La Niña cycles.

**Recommendation 6: EO Solutions for Macroplastic Detection from Space**

As described in chapter 4, a solution to detect aggregated floating plastic debris in oceans is a space- based EO approach that Topouzelis, Papakonstantinou and Garaba (2019) and Biermann, et al. (2020) have studied. The approach suggests using reflectance spectroscopy and SAR working in tandem to scan the ocean.

This report followed the guidance of Biermann, et al. (2020) to test the techniques and found that locally noise interferes with the spectral reflection. The reflection makes it very difficult to differentiate plastic pollution from other things floating on the ocean, such as small boats. In addition to this, the geographical conditions, like shorelines, near the observed location sometimes amplified the reflection, leading to misinterpretations of the data and other substances like non-photosynthetic plants, present a similar chemical structure. They are hence capable of interfering, and it is not easy to separate their spectral signal from the spectral bandwidth of plastic.

Possible solutions include an oversampling mechanism and sampling scenarios. The samples of oceanic plastic trash in target scenarios, like Biermann, et al. (2020) did, can be aggregated and properly processed by using oversampling technique. This way, the images' resolution can be potentially reconstructed, and the detection hence more precise. Research on the implementation of oversampling is, consequently, recommended. Alternatively, the hyperspectral sensors' optimization can be researched to improve the signal-to-noise ratio (SNR) across the spectra.

To overcome difficulties in differentiating artificial objects and contrasting plastic from vegetation, SAR through the reflection in NIR and SWIR, could be used. This leads to the introduction of the NDVI optimizations that can easily identify green vegetation, further explained in section 4.3.1. The authors discovered green vegetation's aggregation during the previously mentioned tests near floating debris patches. These aggregations are most likely algae and other green organisms. One downside to Biermann's strategy is that it relies on analyzing every single pixel. Even with the help of an AI, it would cost immense computational resources, time and money.

It is recommended to conduct further studies regarding the possible correlation between plastic debris patches and green vegetation accumulation in waters. Using the NDVI method, the authors propose examining a photo of the area and performing the NDVI analysis to detect pollution patch-candidates. This way, the NDVI is an implicit marker of plastic patches. Following detection, isolate the patch-candidates and finally, use an AI algorithm to confirm plastic debris.

### Recommendation 7: High-altitude Balloons

As mentioned in the previous recommendation, stratospheric balloons can complement satellite observations. They can continuously observe one area as opposed to satellite observation, which is dependent on revisit times and duration. The stratollites could coordinate AUV missions when no GPS reception is possible and help with synchronizations between the dives and the orbits. Moreover, they have the advantage of producing less noise and pollution compared to other vehicles and can be launched from virtually anywhere, including near inhabited areas. To a certain degree, balloons are weather dependent but since they are only in the air for a short amount of time, it is a manageable limitation. Furthermore, they need to be collected once they land to avoid contributing to environmental pollution; therefore, the authors encourage further investigation of the life expectancy of these balloons and possible end-of-life scenarios.

### Recommendation 8: An Integrated Solution

Based on the project's research, an integrated and multilevel plan for the detection of plastic pollution was presented in section 4.6, using technologies such as AUVs and stratospheric balloons to compensate when EO from space is unavailable or low-quality. The recommendation of this paper is to integrate ground-based technologies with space-based ones to maximise efficiency. For example, one could use space-based technology to detect and track plastic and subsequently downlink such information about significant accumulation sites' locations to in-situ cleanup technology. Live-tracking of ocean plastic movement could be provided in an open-source way that allows projects such as the Great Bubble Barrier and Good Karma Projects to set up their mitigation and cleanup efforts at the most impacted locations. Additionally, an open-source database of both live and historical ocean plastic movement and accumulation patterns could be created. A database like this could inform nations within these regions to work toward trans-national solutions. The UN could sponsor this to happen and facilitate its adoption by member states who have already adopted 'green' practices regarding plastic pollution. In this way, an international consortium could form and specifically address ocean plastic pollution on national, regional, and global scales.

Due to the scope of the project, solutions to preventing plastic pollution other than monitoring and tracking were not considered but as pointed out by Sonja Behlmel, CEO and Founder of WaterShed Monitoring, the data accumulated from tracking methods can be immediately used by interested groups to prevent plastic pollution. Therefore, it is recommended to investigate the different strategy component interdependencies and build a mission and a business plan for the suggested solution.

### 6.2.3 Science Diplomacy and Outreach Recommendation 9: Government Level

As decision makers with the ability to enact policies and mobilise change, it is very important to get the government level to understand how plastic pollution affects all ecosystems and requires a coordinated effort to tackle. As for passing new legislation on plastic pollution, there are four things to consider a) there is no international agreed upon scale of the marine plastic pollution problem, b) each region impacted has different priorities, c) some governments are prone to be more influenced by lobbyists from the petrochemical industry, resulting in no action or harmful action, and d) environmental policies tend not to be bipartisan issues, facing resistance from counterparts. It is recommended to have a neutral outreach program that can be used for every country and meets bipartisan goals. That way, it is less likely to provide a biased message and more likely to be adopted. Furthermore, it is suggested that politicians provide a clear and realistic plan and solution when they want to use marine plastic pollution on their agenda and are held accountable.

**Recommendation 10: Public Level**

The second audience for providing science and policy education is the general public. Through the use of hashtags and social media, information can be disseminated about plastic pollution on multiple platforms. However, if not chosen carefully, the hashtag can be improperly used or adopted by other interest groups to polarise the content creating greater problems. To avoid misuse of the hashtag authors recommend to keep it simple and to combine the hashtag with scientific and relevant content such as a plastic campaign that can be shared by research institutions, individuals and groups.

## 6.3 Final Statement

This report has covered the necessity of monitoring plastic pollution, specifically in the Arctic Ocean, whilst highlighting the technological, economical and political challenges that will be faced. The problem statement which directed our research was 'How can we monitor ocean plastic pollution?'. Henceforth, the solutions based on space technologies proposed were centered around the Arctic Ocean with the aim to scale it to other oceans as this is a global issue. Ultimately, we have found that no one solution, be it drones, satellites or other, is sufficient in itself for an effective and consistent monitoring of all plastics. An integration of separate technologies and approaches will be necessary if the problem is to be assessed and, eventually, resolved. Our findings allow us to conclude that space should play a role in this complex and articulated combination of technologies. It offers physical and operational advantages that are not only useful but unique, and would therefore play a key role in the solution to this worldwide problem.

However, substantial governmental efforts must be made to implement such directives. Cost regulation is also key in the application of space based technologies. Public activity, such as the SpaceCleaning initiative proposed in section 5.4, and education is crucial in spreading global awareness of the current state of plastic within the Arctic Ocean. Hereby with this report, the authors intend to bring attention to the usefulness and necessity of creative uses of space applications complemented with innovative technologies from different sectors, as a means to monitor ocean plastic debris. We expect that our solutions will encourage researchers to investigate using a combination of technologies associated with space, with the sole aim of solving plastic pollution at a global level.

# References

5Gyres.Org, 2020. *Plastic FAQs.* [online]. Available at: <www.5gyres.org> [Accessed March 3, 2021].

Abalansa, S, El Mahrad, B, Vondolia,GK, Icely, J and Newton, A, 2020. The Marine Plastic Litter Issue: A Social–Economic Analysis. *Sustainability,* [e–journal] 12(20), 8677. http://dx.doi.org/10.3390/su12208677.

Abrams, M, 2019. *A Robot Crab to Clean the Ocean.* [online]. American Society of Mechanical Engineers (ASME). Available at: <https://www. asme.org/topics–resources/content/a– robot–crab–to–clean–the–ocean> [Accessed March 3, 2021].

Abreu, A. and Pedrotti, ML. editors, 2019. *Microplastics in the oceans: the solutions lie on land.* Volume [e–journal] Special Issue 19, 62–67.

Adobe, 2021. *Arctic Ocean map with North Pole and Arctic Circle.* [online]. Available at: <https: //stock.adobe.com/fr/images/arctic–ocean–map–with–north–pole–and–arctic–circle–arctic–region–map–with–countries–national–borders–rivers–and–lakes–map–without–sea–ice–english– labeling–and–scaling/113564275> [Accessed March 10, 2021].

Agamuthu, P, Mehran, S. B, Norkhairah, A and Norkhairiyah, A, 2019. Marine debris: A review of impacts and global initiatives. *Waste management & research : the journal of the International Solid Wastes and Public Cleansing Association, ISWA,* [e–journal] 37(10). 9871002. http://dx.doi.org/10.1177/0734242X19845041.

Airbus, 2021. *Pléiades Neo Ready for Launch.* [online]. Available at: <https:// www.intelligence– airbusds.com/newsroom/news/pleiades–neo–ready–for–launch/> [Accessed March 3, 2021].

Airbus Defence and Space, 2021. *Pléiades Neo, Trusted Intelligence.* Available at: <https://www. intelligence–airbusds.com/en/8671–pleiades–neo–trusted–intelligence> [Accessed March 3, 2021].

145

Aksenov, Y, Popova, E E, Yool, A, Nurser, AG, Williams, T D, Bertino, L. and Bergh, J, 2017. On the future navigability of Arctic sea routes: High-resolution projections of the Arctic Ocean and sea ice. *Marine Policy,* [e-journal] 75. 300–317. http://dx.doi.org/10. 1016/j.marpol.2015.12.027.

Alpizar, F, Carlsson, F, Lanza, G, Carney, B, Daniels, RC, Jaime, M, Ho, T, Nie, Z, Salazar, C, Tibesigwa, B. and Wahdera, S, 2020. A framework for selecting and designing policies to reduce marine plastic pollution in developing countries. *Environmental Science & Policy,* [e-journal] 109. 25–35. http://dx.doi.org/10.1016/j.envsci.2020.04.007.

Altendorf, D, 2020. *Comprendre la force de Coriolis et son rôle dans la dynamique de l'atmosphère et de l'océan.* [online]. Available at: <https://sciencepost.fr/comprendre-la-force-de-coriolis-et-son-role-dans-la-dynamique-de-latmosphere-et-de-locean/> [Accessed March 3, 2021].

American Geosciences, 2016. *What Are El Nino and La Niña?* [online]. American Geosciences. Available at: <https://www.americangeosciences. org/critical-issues/faq/what-are-el-nino- and-la-nina>.

Andronov, S, Lobanov, A, Popov, A, Luo, Y, Shaduyko, O, Fesyun, A, Lobanova, L, Bogdanova, E. and Kobel'kova, I, 2020. Changing diets and traditional lifestyle of Siberian Arctic Indigenous Peoples and effects on health and well-being. *Ambio,* [online]. http: //dx.doi.org/10.1007/ s13280-020-01387-9.

Antoni, A, Back, M, Bauraing, G, Baustert, L, Clemens, A, Diels, A, Divo, L, Estgen, L, Ettinger Gregory, Garçon, J, Hang Hélène, Hinh, A, Hoscheid, Y, Kloeckner, C, Marx, E, Mathieu, L, Nau, C, Ni, J, Nguyen, J, Schaeffner, C, Schaus, V, Toni, L, Wantz, L, Wild, C, Weber, X, Philipps, M–F. and Wilwert, C, 2020. fish or plastic? [exhibition] (Luxembourg, February 21, 2021).

Anwar, S, 2017. *What Are the Differences between El Nino and La Nina?* [online]. Available at: <https://www.jagranjosh.com/general-knowledge/what-are-the-differences-between-el- nino-and-la-nina-1511951289-1> [Accessed March 3, 2021].

Arctic Council Secretariat, 2021. *How We Work.* [online]. Available at: <https://arctic-council. org/en/explore/work/> [Accessed March 3, 2021].

Arctic Portal, 2019. *Alarming visualization of changing sea ice in the Arctic over the past 35 years - NASA.* [online]. Available at: <https://arcticportal. org/ap-library/news/2204-alarming- visualization-of-changing-sea-ice-in-the-arctic-over-the-past-35-years-nasa> [Accessed March 10, 2021].

Armitage, T. W. K, Manucharyan, GE, Petty, AA, Kwok, R. and Thompson, AF, 2020. Enhanced eddy activity in the Beaufort Gyre in response to sea ice loss. *Nature communications,* [e-journal] 11(1), 761. http://dx.doi.org/10.1038/s41467-020-14449-z.

Ashbullby, KJ, Pahl, S, Webley, P, and White, MP, 2013. The beach as a setting for families' health promotion: a qualitative study with parents and children living in coastal regions in Southwest England. *Health & Place,* [e-journal] 23. 138-147. http://dx.doi.org/10.1016/j. healthplace.2013.06.005.

Awuchi, CG. and Awuchi, CG, 2019. Impacts of Plastic Pollution on the Sustainability of Seafood Value Chain and Human Health. *International Journal of Advanced Academic Research,* [e-journal] 5(11). 46-138. Available at: <https://www.researchgate.net/publication/ 337312788_ Impacts_of_Plastic_Pollution_on_the_Sustainability_of_Seafood_ Value_Chain_ and_Human_Health> [Accessed March 3, 2021].

Barnes, DKA, Galgani, F, Thompson, RC and Barlaz, M, 2009. Accumulation and fragmentation of plastic debris in global environments. *Philosophical transactions of the Royal Society of London. Series B, Biological sciences,* [e-journal] 364(1526). 1985-1998. http://dx.doi.org/10.1098/rstb.2008.0205.

BBC, 2019. *MSC Zoe: Islands Hit as 270 Containers Fall off Ship.* [online]. BBC News. Available at: <https://www.bbc.com/news/world-europe-46746312> [Accessed March 3, 2021].

Beaumont, NJ, Aanesen, M, Austen, MC, Börger, T, Clark, JR, Cole, M, Hooper, T, Lindeque, K, Pascoe, C. and Wyles, KJ, 2019. Global ecological, social and economic impacts of marine plastic. *Marine Pollution Bulletin,* [e-journal] 142. 189-195. http: //dx.doi.org/10.1016/j. marpolbul.2019.03.022.

Bell, J, Nel, and Stuart, B, 2019. Non-invasive identification of polymers in cultural heritage collections: evaluation, optimisation and application of portable FTIR (ATR and external reflectance) spectroscopy to three-dimensional polymer-based objects. *Heritage Science,* [e-journal] 7(1). http://dx.doi.org/10.1186/s40494-019-0336-0.

Bentley, J, 2021. Detecting Ocean Microplastics with Remote Sensing in the Near-Infrared: A Feasibility Study. In: *BSU Honors Program Theses and Projects.* Volume [e-journal] Item 309. Available at: <https://vc.bridgew.edu/honors_proj/309> [Accessed March 3, 2021].

Bergmann, M, Mützel, S, Primpke, S, Tekman, MB, Trachsel, J. and Gerdts, G, 2019. White and wonderful? Microplastics prevail in snow from the Alps to the Arctic. *Science Advances,* [e-journal] 5(8), eaax1157. http://dx.doi.org/10.1126/sciadv.aax1157.

Bergmann, M, Sandhop, N, Schewe, I. and D'Hert, D, 2016. Observations of floating anthropogenic litter in the Barents Sea and Fram Strait, Arctic. *Polar Biology*, [e-journal] 39(3). 553–560. http://dx.doi.org/10.1007/s00300–015–1795–8.

Biermann, L, Clewley, D, Martinez–Vicente, V and Topouzelis, K, 2020. Finding Plastic Patches in Coastal Waters using Optical Satellite Data. *Scientific Reports*, [e-journal] 10(1), 5364. http://dx.doi.org/10.1038/s41598–020–62298–z.

Binder, C, 2016. *Science as Catalyst for Deeper Arctic Cooperation? Science Diplomacy and the Transformation of the Arctic Council.* [online]. Available at: <https://arcticyearbook.com/arctic-yearbook/2016/2016–scholarly-papers/171–science-as-catalyst-for-deeper-arctic-  cooperation-science-diplomacy-and-the-transformation-of-the-arctic-council> [Accessed March 3, 2021].

Bjerregaard, P, Kue Young, T, Dewailly, E. and Ebbesson, SO, 2004. Review Article: Indigenous Health in the Arctic: An Overview of the Circumpolar Inuit Population. *Scandinavian Journal of Public Health*, [e-journal] 32(5). 390–395. http:// dx.doi.org/10.1080/14034940410028398.

Björnsson, H. and Pálsson, F, January 2008. Icelandic glaciers. *Jökull*, [e-journal] 58.

Bogden, S. and Edwards, CA, 2001. Wind Driven Circulation. In: JH. Steele, KK Turekian and SA Thorpe editors. *Encyclopedia of Ocean Sciences.* Amsterdam: Elsevier. 3227–3236. http://dx.doi.org/10.1006/rwos.2001.0110.. Available at: <https://www.sciencedirect.com/ topics/earth–and–planetary–sciences/wind–driven–circulation> [Accessed March 3, 2021].

Boucher, J and Billard, G, 2020. *The Mediterranean: Mare Plasticum.* [online]. Available at: <https://portals.iucn.org/library/sites/library/files / documents/2020–030–En.pdf> [Accessed March 3, 2021].

Bradley, M, 2021. *FTIR Sample Techniques – True Specular Reflectance/ Reflection–Absorption – US.* [online]. Available at: <//www.thermofisher.com/us/en/home/industrial/spectroscopy- elemental – isotope – analysis / spectroscopy – elemental – isotope – analysis – learning – center / molecular-spectroscopy-information/ftir-information/ftir–sample-handling-techniques/ftir–sample-handling-techniques-true-specular-reflectance-reflection–absorption.html> [Accessed March 3, 2021].

Bramston, P, Pretty, G. and Zammit, C, 2011. Assessing Environmental Stewardship Motivation. *Environment and Behavior*, [e-journal] 43(6). 776–788. http://dx.doi.org/10.1177/ 0013916510382875.

British Antarctic Survey, 2020. *Past Evidence Supports Complete Loss of Arctic Sea-Ice by 2035.* [online]. Science Daily. Available at: <https://www.sciencedaily.com/releases/2020/08/ 200810113216.htm> [Accessed March 3, 2021].

Brouwer, R, Hadzhiyska, D, Ioakeimidis, C and Ouderdorp, H, 2017. The social costs of marine litter along European coasts. *Ocean & Coastal Management,* [e-journal] 138. 38–49. http://dx.doi.org/10.1016/j. Ocecoaman.2017.01.011.

Bundesministerium für Bildung und Forschung, 2018. *Who lives in the Arctic? - BMBF Arctic Science Ministerial.* [online]. Available at: <https://www.arcticscienceministerial.org/en/who-lives-in-the-arctic-1731.html> [Accessed March 10, 2021].

Canadian Space Agency, 2021. *RADARSAT-2.* Available at: <https://www.asc-csa.gc.ca/eng/ satellites/radarsat2/Default.asp> [Accessed March 3, 2021].

Carr, S, 2019. *What Is Marine Plastic Pollution Costing Us? The Impacts of Marine Plastic on the Blue Economy.* [online]. The Skimmer. Available at: <https://meam.openchannels.org/news/skimmer-marine-ecosystems-and-management/what-marine-plastic-pollution-costing-us-impacts> [Accessed March 8, 2021].

Chidepatil, A, Bindra, P, Kulkarni, D, Qazi, M, Kshirsagar, M and Sankaran, K, 2020. From Trash to Cash: How Blockchain and Multi-Sensor-Driven Artificial Intelligence Can Transform Circular Economy of Plastic Waste? *Administrative Sciences,* [e-journal] 10(2), 23. http://dx.doi.org/10.3390/admsci10020023.

Cho, D-J and Kim, K-Y, 2021. Role of Ural blocking in Arctic sea ice loss and its connection with Arctic warming in winter. *Climate Dynamics,* [e-journal] 56(5–6). 1571–1588. http://dx.doi.org/10.1007/s00382-020-05545-3.

Cisneros-Montemayor, AM, Pauly, D, Weatherdon, LV and Ota, Y, 2016. A Global Estimate of Seafood Consumption by Coastal Indigenous Peoples. *PLOS ONE,* [e-journal] 11(12), e0166681. http://dx.doi.org/10.1371/journal.pone.0166681.

Clear Blue Sea, 2020. *Meet Fred Clear Blue Sea.*

Coll, M, Piroddi, C, Steenbeek, J, Kaschner, K, Ben Rais Lasram, F, Aguzzi, J, Ballesteros, E, Bianchi, CN, Corbera, J, Dailianis, T, Danovaro, R, Estrada, M, Froglia, C, Galil, B.S, Gasol, JM, Gertwagen, R, Gil, J, Guilhaumon, F, Kesner-Reyes, K, Kitsos, M.-S, Koukouras, A, Lampadariou, N, Laxamana, E, La López-Fé de Cuadra, CM, Lotze, HK, Martin, D, Mouillot, D, Oro, D, Raicevich, S, Rius-Barile, J, Saiz-Salinas, JI, San Vicente, C, Somot,

S, Templado, J, Turon, X, Vafidis, D, Villanueva, R. and Voultsiadou, E, 2010. The biodiversity of the Mediterranean Sea: estimates, patterns, and threats. *PLOS ONE*, [e–journal] 5(8), e11842. http://dx.doi.org/10.1371/journal.pone.0011842.

Comiso, JC, 2012. Large Decadal Decline of the Arctic Multiyear Ice Cover. *Journal of Climate*, [e–journal] 25(4). 1176–1193. http://dx.doi.org/10.1175/JCLI-D-11-00113.1.

Cooke, SJ, 2016. How do oceans affect weather and climate? I Socratic. *Socratic.org*, [blog] 12 March Available at: <https://socratic.org/questions/how–do–oceans–affect–weather–and– climate> [Accessed March 3, 2021].

Copeland, BJ, 2020. Artificial Intelligence Definition, Examples, and Applications. Available at: <https://www.britannica.com/technology/artificial–intelligence> [Accessed March 3, 2021].

Copernicus, 2021. *From Plastic to Marine Pollution: CMEMS*. [online]. Available at: <https://marine. copernicus.eu/services/plastic–pollution/from–plastic–marine–pollution> [Accessed March 3, 2021].

Cordier, M and Uehara, T, 2019. How much innovation is needed to protect the ocean from plastic contamination? *The Science of the total environment*, [e–journal] 670. 789–799. http://dx.doi.org/10.1016/j.scitotenv.2019.03.258.

Council on Foreign Relations, 2021. *The Emerging Arctic*. [online]. Available at: <http://www. cfr.org/arctic> [Accessed March 3, 2021].

Cózar, A, Martí, E, Duarte, CM, García–de–Lomas, J, van Sebille, E, Ballatore, TJ, Eguíluz, VM, González–Gordillo, JI, Pedrotti, ML, Echevarría, F, Troublè, R and Irigoien, X, 2017. The Arctic Ocean as a dead end for floating plastics in the North Atlantic branch of the Thermohaline Circulation. *Science Advances*, [e–journal] 3(4), e1600582. http://dx.doi.org/10.1126/sciadv.1600582.

Cózar, A, Sanz–Martín, M, Martí, E, González–Gordillo, JI, Ubeda, B, Galvez, JÁ, Irigoien, X and Duarte, CM, 2015. Plastic accumulation in the Mediterranean sea. *PLOS ONE*, [e–journal] 10(4), e0121762. http://dx.doi.org/10.1371/journal.pone.0121762.

Craig, K, 2016. *The Arctic Frontier in International Relations*. [online]. Available at: <https://www. researchgate.net/publication/331547988%5Ctextunderscore%20The%5Ctextunderscore% 20Arctic%5Ctextunderscore%20Frontier%5Ctextunderscore% 20in%5Ctextunderscore%20International%5Ctextunderscore%20Relations> [Accessed March 3, 2021].

Daly, S, 2008. Evolution of Frazil ice using new technologies to understand water–ice interaction. In: *Proc. 19th IAHR Int. Symp. on Ice*. 29–50.

Dassault Systèmes, 2021. *ZERO 2 INFINITY.* [online]. Available at: <https://3dexperiencelab. 3ds.com/en/projects/city/zero-2-infinity/> [Accessed March 3, 2021].

Dauvergne, P, 2018. Why is the global governance of plastic failing the oceans? *Global Environmental Change,* [e-journal] 51. 22-31. http:// dx.doi.org/10.1016/j.gloenvcha.2018.05.002.

Dekker, AG. and Hestir, EL, 2012. *Evaluating the Feasibility of Systematic Inland Water Quality Monitoring with Satellite Remote Sensing.* [online]. Available at: <https://publications.csiro. au/rpr/download?pid=csiro:EP1 17441&dsid=DS10> [Accessed March 3, 2021].

Deloitte, 2019. *The Price Tag of Plastic Pollution An Economic Assessment of River Plastic.* [online]. Available at: <https://www2.deloitte.com/content/ dam/Deloitte/nl/Documents/strategy-analytics-and-ma/deloitte-nl-strategy-analytics-and-ma-the-price-tag-of-plastic-pollution.pdf> [Accessed March 3, 2021].

Deutsches Zentrum für Luft- und Raumfahrt, 2021. *EnMAP: Earth Observation Center.* [online]. DLR. Available at: <https://www.dlr.de/eoc/ en/desktopdefault.aspx/tabid-5514/20470% 5Ctextunderscore%20read-47899/> [Accessed March 3, 2021].

Dodds, K, 2013. The Ilulissat Declaration (2008): The Arctic States, 'Law of the Sea,' and Arctic Ocean. *SAIS Review of International Affairs,* [e-journal] 33(2). 45-55. http://dx.doi.org/10. 1353/sais.2013.0018.

Doyle, A, 2018. Plastic Waste in Antarctica Reveals Scale of Global Pollution: Greenpeace. *Reuters.* Available at: <https ://www.reuters.com/article/us – antarctica-plastics-idUSKCN1J22YW> [Accessed March 3, 2021].

Dris, R, Gasperi, J, Rocher, V, Saad, M, Renault, N. and Tassin, B, 2015. Microplastic contamination in an urban area: a case study in Greater Paris. *Environmental Chemistry,* [e-journal] 12(5), 592. http://dx.doi. org/10.1071/EN14167.

Duncan, EM, Arrowsmith, J, Bain, C, Broderick, AC, Lee, J, Metcalfe, K, Pikesley, SK, Snape, RTE, van Sebille, E and Godley, BJ, 2018. The true depth of the Mediterranean plastic problem: Extreme microplastic pollution on marine turtle nesting beaches in Cyprus. *Marine Pollution Bulletin,* [e-journal] 136. 334-340. http://dx.doi.org/10.1016/j. marpolbul.2018.09.019.

Dupeyroux, J, Viollet, S. and Serres, JR, 2019. An ant-inspired celestial compass applied to autonomous outdoor robot navigation. *Robotics and Autonomous Systems,* [e-journal] 117. 40-56. http://dx.doi.org/10.1016/j. robot.2019.04.007.

Eastman, LB, Núñez, P, Crettier, B. and Thiel, M, 2013. Identification of self-reported user behavior, education level, and preferences to reduce littering on beaches – A survey from the SE Pacific. *Ocean & Coastal Management,* [e-journal] 78. 18–24. http://dx.doi.org/10. 1016/j. ocecoaman.2013.02.014.

Ellen Macarthur Foundation, 2017. *The New Plastics Economy: Rethinking the Future of Plastics & Catalysing Action.* [online]. Available at: <https://www. ellenmacarthurfoundation.org/ assets/downloads/publications/NPEC-Hybrid_English_22-11-17_Digital.pdf> [Accessed March 3, 2021].

Emmerik, T. and Schwarz, A, 2020. Plastic debris in rivers. *WIREs Water,* [e-journal] 7(1). http://dx.doi.org/10.1002/wat2.1398.

EnMAP Ground Segment Team, 2020. *EnMAP – Spaceborne Imaging Spectroscopy Mission Compilation.*

Environmental Investigation Agency, 2020. *Plastic Pollution from Ships EIA Reports.* [online]. EIA International. Available at: <https://reports. eia-international.org/globalplastics/plastic- pollution-from-ships/> [Accessed March 3, 2021].

Eriksen, M, Lebreton, L, Carson, HS, Thiel, M, Moore, CJ, Borerro, JC, Galgani, F, Ryan, G and Reisser, J, 2014. Plastic Pollution in the World's Oceans: More than 5 Trillion Plastic Pieces Weighing over 250,000 Tons Afloat at Sea. *PLOS ONE,* [e-journal] 9(12), e111913. http://dx.doi. org/10.1371/journal.pone.0111913.

European Chemicals Agency, 2020. *Microplastics.* [online]. Available at: <https://echa.europa. eu/en/hot-topics/microplastics> [Accessed March 3, 2021].

European Space Agency, 2008. *TerraSAR-X and TanDEM-X.* [online]. Available at: <http://www.esa.int/ESA%5Ctextunderscore%20 Multimedia/Images/2011/03/TerraSAR-X%5Ctextunderscore%20 and%5Ctextunderscore%20TanDEM-X> [Accessed March 3, 2021].

European Space Agency, 2017. *EnMAP (Environmental Monitoring and Analysis Program).* [online]. Available at: <https://directory.eoportal.org/ web/eoportal/satellite-missions/e/enmap> [Accessed March 3, 2021].

European Space Agency, 2018. *Bloostar Microlauncher.* Available at: <https://www.esa.int/ESA_Multimedia/Images/2018/02/Bloostar_ microlauncher>.

European Space Agency, 2019a. *Copernicus Sentinel-2 Improves Observations of Lakes and Water Bodies – Sentinel.* [online]. Available at: <https: // sentinels.copernicus eu/web/sentinel/news/success-stories/-/asset_ publisher/3H6l2SEVD9Fc/content/copernicus-sentinel-2-improves-observations-of-lakes-and-water-bodies;jsessionid=AC4D716EDC5

A1C043811D6AE141FCED1.jvm1?redirect=https%3A%2F%2Fsentin els.copernicus.eu%2Fweb%2Fsentinel%2Fnews % 2Fsuccess - stories % 3Bjsessionid % 3DAC4D716EDC5A1C043811D6AE141FCED1. jvm1 %3Fp_p_id%3D101_INSTANCE_3H6l2SEVD9Fc%26p_p_ lifecycle%3D0%26p_p_state%3Dnormal%26p_p_mode%3Dview% 26p_p_col_id%3Dcolumn-1%26p_p_col_pos%3D1%26p_p_col_ count%3D2> [Accessed March 3, 2021].

European Space Agency, 2019b. Seeking Innovative Ideas: Space for the Oceans. *The European Space Agency,* [online]. Available at: <https://www. esa.int/Enabling_Support/Space_Engineering_Technology/Seeking_ innovative_ideas_space_for_the_oceans> [Accessed March 3, 2021].

European Space Agency, 2019c. *Vega Launches PRISMA for Italy.* [online]. Available at: <https://www.esa.int/Enabling_Support/Space_ Transportation/Vega_launches_PRISMA_for_Italy> [Accessed March 3, 2021].

European Space Agency, 2020a. *PRISMA (Hyperspectral).* [online]. Available at: <https: //directory.eoportal.org/web/eoportal/satellite-missions/p/ prisma-hyperspectral> [Accessed March 3, 2021].

European Space Agency, 2020b. *Radarsat-2.* [online]. Available at: <https:// directory.eoportal. org/web/eoportal/satellite-missions/r/radarsat-2> [Accessed March 3, 2021].

European Space Agency, 2020c. *TanDEM-X.* [online]. Available at: <https:// spacedata. copernicus.eu/data-offer/missions/tandem-x> [Accessed March 3, 2021].

European Space Agency, 2021a. *Copernicus Open Access Hub.* Available at: <https://scihub. copernicus.eu/dhus/#/home>.

European Space Agency, 2021b. *Sentinel-1 - Instrument Payload - Sentinel Online - Sentinel.* [online]. Available at: <https://sentinel.esa.int/web/ sentinel/missions/sentinel-1/instrument-payload> [Accessed March 3, 2021].

European Space Agency, 2021c. *Sentinel-2.* [online]. Available at: <https:// sentinels.copernicus. eu/web/sentinel/missions/sentinel-2> [Accessed March 3, 2021].

European Space Agency, 2021d. *Sentinel-3 - Overview - Sentinel Online.* [online]. Available at: <https://sentinel.esa.int/web/sentinel/missions/ sentinel-3/overview> [Accessed March 3, 2021].

European Space Agency, 2021e. *User Guides - Sentinel-2 MSI - Revisit and Coverage - Sentinel Online - Sentinel.* [online]. Available at: <https:// sentinels.copernicus.eu/web/sentinel/user- guides/sentinel-2-msi/ revisit-coverage> [Accessed March 3, 2021].

Evers, J and Editing, E, 2019. *Great Pacific Garbage Patch.* Edited by NGS. Caryl–Sue. [online]. Available at: <http://www.nationalgeographic.org/encyclopedia/great–pacific–garbage– patch/> [Accessed March 3, 2021].

Exner–Pirot, H, Maria Ackrén, Natalia Loukacheva, Heather Nicol, Annika E. Nilsson and Jennifer Spence, 2019. *Form and Function: The Future of the Arctic Council.* [online]. Available at: <https://www.thearcticinstitute.org/form–function–future–arctic–council/> [Accessed March 3, 2021].

Folger, T, 2017. *Arctic Wildlife's Last Habitat Will Be Ice Strip.* [online]. Available at: <https: //www.nationalgeographic.com/magazine/article/arctic–wildlife–sea–ice> [Accessed March 3, 2021].

Fondahl, G, Filippova, V and Mack, L, 2015. Indigenous Peoples in the New Arctic. In: B Evengard, J Nymand Larsen and 0 Paasche editors, *The New Arctic.* Cham: Springer International Publishing. 7–22. **isbn:** 978–3–319–17601–7. http://dx.doi.org/10.1007/978– 3–319–17602–4{\textunderscore}2.

Foust, J, 2020. *World View Founders Launch New Stratospheric Ballooning Venture.* [online]. Space News. Available at: <https://spacenews.com/world–view–founders–launch–new–stratospheric–ballooning–venture/> [Accessed March 3, 2021].

Fowler, MC, Bolding, TL, Hebert, KM, Ducrest, F and Kumar, A, 2016. Design of a cost–effective autonomous underwater vehicle. In: *2016 Annual IEEE Systems Conference (SysCon).* Volume [pdf]. 1–6. http://dx.doi.org/10.1109/SYSCON.2016.7490543.

Fretter, H, 2017. *Could a floating shipping container sink your yacht? How real is the danger?* [online]. Yachting World. Available at: <https://www.yachtingworld.com/news/could–a–floating– shipping–container–sink–your–yacht–is–the–danger–to–sailors–real–or–imagined–107508> [Accessed March 3, 2021].

Fulton, M, Hong, J, Islam, MJ. and Sattar, J, 2018. Robotic Detection of Marine Litter Using Deep Visual Detection Models. *2019 International Conference on Robotics and Automation (ICRA).* 5752–5758. http://dx.doi.org/10.1109/ICRA.2019.8793975.

Galley, RJ, Else, BG, Geilfus, N.-X, Hare, AA, Babb, D, Papakyriakou, T, Barber, DG. and Rysgaard, S, 2015. Micrometeorological and Thermal Control of Frost Flower Growth and Decay on Young Sea Ice. *ARCTIC,* [e–journal] 68(1), 79. http://dx.doi.org/10.14430/ arctic4457.

Garaba, SP, Aitken, J, Slat, B, Dierssen, HM, Lebreton, L, Zielinski, O. and Reisser, J, 2018. Sensing Ocean Plastics with an Airborne Hyperspectral Shortwave Infrared Imager. *Environmental Science & Technology,* [e–journal] 52(20). 11699–11707. http://dx.doi.org/ 10.1021/acs.est.8b02855.

Garaba, SP, Arias, M, Corradi, P, Harmel, T, Vries, R de and Lebreton, L, 2021. Concentration, anisotropic and apparent colour effects on optical reflectance properties of virgin and ocean-harvested plastics. *Journal of Hazardous Materials*, [e-journal] 406, 124290. http://dx.doi.org/10.1016/j. jhazmat.2020.124290.

Gasperi, J, Dris, R, Bonin, T, Rocher, V and Tassin, B, 2014. Assessment of floating plastic debris in surface water along the Seine River. *Environmental pollution (Barking, Essex : 1987)*, [e-journal] 195. 163–166. http://dx.doi. org/10.1016/j.envpol.2014.09.001.

Geilfus, N–X, Munson, KM, Sousa, J, Germanov, Y, Bhugaloo, S, Babb, D and Wang, F, 2019. Distribution and impacts of microplastic incorporation within sea ice. *Marine Pollution Bulletin*, [e-journal] 145. 463–473. http:// dx.doi.org/10.1016/j.marpolbul.2019.06.029. [Accessed March 3, 2021].

Geyer, R, Jambeck, JR. and Law, KL, 2017. Production, use, and fate of all plastics ever made. *Science Advances*, [e-journal] 3(7), e1700782. http:// dx.doi.org/10.1126/sciadv.1700782. Gibbons, S, 2019. *Another Plastic Bag Found at the Bottom of World's Deepest Ocean Trench*. [online]. Available at: <https://www.nationalgeographic.org/article/plastic-bag-found-bottom- worlds-deepest-ocean-trench/> [Accessed March 3, 2021].

Goddijn–Murphy, L, Peters, S, van Sebille, F, James, NA. and Gibb, S, 2018. Concept for a hyperspectral remote sensing algorithm for floating marine macro plastics. *Marine Pollution Bulletin*, [e-journal] 126. 255–262. http://dx.doi.org/10.1016/j.marpolbul.2017.11.011. González–García, J, Gómez–Espinosa, A, Cuan–Urquizo, E, García–Valdovinos, LG, Salgado– Jiménez, T. and Cabello, JAE, 2020. Autonomous Underwater Vehicles: Localization, Navigation, and Communication for Collaborative Missions. *Applied Sciences*, [e-journal] 10(4), 1256. http://dx.doi. org/10.3390/app10041256.

Gosnell, R, 2018. *The Complexities of Arctic Maritime Traffic*. [online]. Available at: <https://www.thearcticinstitute.org/complexities–arctic–maritime–traffic/> [Accessed March 3, 2021].

Gove, JM, Whitney, JL, McManus, MA, Lecky, J, Carvalho, FC, Lynch, J. M, Li, J, Neubauer, P, Smith, KA, Phipps, JE, Kobayashi, DR, Balagso, KB, Contreras, EA, Manuel, ME, Merrifield, MA, Polovina, JJ, Asner, GP, Maynard, JA. and Williams, GJ, 2019. Prey–size plastics are invading larval fish nurseries. *Proceedings of the National Academy of Sciences*, [e-journal] 116(48). 24143–24149. http://dx.doi.org/10.1073/pnas. 1907496116.

GRID Arendal, 2007. *World Ocean Thermohaline Circulation*. [online]. Available at: <https://www.grida.no/resources/5228> [Accessed March 3, 2021].

Guo, X. and Wang, J, 2019. Sorption of antibiotics onto aged microplastics in freshwater and seawater. *Marine Pollution Bulletin,* [e-journal] 149, 110511. http://dx.doi.org/10.1016/j.marpolbul.2019.110511.

Güven, O, Gökda, K, Jovanovi, B and Kdey, AE, 2017. Microplastic litter composition of the Turkish territorial waters of the Mediterranean Sea, and its occurrence in the gastrointestinal tract of fish. *Environmental pollution (Barking, Essex : 1987),* [e-journal] 223. 286–294. http://dx.doi.org/10.1016/j.envpol.2017.01.025.

Halsband, C. and Herzke, D, 2019. Plastic litter in the European Arctic: What do we know? *Emerging Contaminants,* [e-journal] 5. 308–318. http://dx.doi.org/10.1016/j.emcon.2019. 11.001.

He, M, Hu, Y, Chen, N, Wang, D, Huang, J. and Stamnes, K, 2019. High cloud coverage over melted areas dominates the impact of clouds on the albedo feedback in the Arctic. *Scientific Reports,* [e-journal] 9(1), 9529. http://dx.doi.org/10.1038/s41598-019-44155-w.

Heidbreder, L. M, Bablok, I, Drews, S. and Menzel, C, 2019. Tackling the plastic problem: A review on perceptions, behaviors, and interventions. *The Science of the total environment,* [e-journal] 668. 1077–1093. http://dx.doi.org/10.1016/j.Scitotenv.2019.02.437.

Henderson, L, 2019. *Sociological Perspectives on Plastic Pollution.* [online]. Available at: <https://www.britsoc.co.uk/about/latest-news/2019/march/sociological-perspectives-on-plastic- pollution/> [Accessed March 3, 2021].

Hernandez, L. M, Xu, E. G, Larsson, HCE, Tahara, R, Maisuria, VB. and Tufenkji, N, 2019. Plastic Teabags Release Billions of Microparticles and Nanoparticles into Tea. *Environmental Science & Technology,* [e-journal] 53(21). 12300–12310. http://dx.doi.org/10.1021/acs.est. 9b02540.

Herrmann, V, 2017. *An Arctic Council In Transition.* Available at: <https://www.thearcticinstitute.org/arctic-council-in-transition/>.

Hood, M, 2019. Arctic sea ice loaded with microplastics. *Phys.org,* [blog] 16 August. Available at: <https://phys.org/news/2019-08-arctic-sea-ice-microplastics.html> [Accessed March 3, 2021].

Horton, AA, Walton, A, Spurgeon, DJ, Lahive, E and Svendsen, C, 2017. Microplastics in freshwater and terrestrial environments: Evaluating the current understanding to identify the knowledge gaps and future research priorities. *The Science of the total environment,* [e-journal] 586. 127–141. http://dx.doi.org/10.1016/j.scitotenv.2017.01.190.

Howell, EA, Bograd, SJ, Morishige, C, Seki, MP. and Polovina, J. J, 2012. On North Pacific circulation and associated marine debris concentration. *Marine Pollution Bulletin,* [e-journal] 6(1–3). 16–22. http://dx.doi.org/10.1016/j.marpolbul.2011.04.034.

Hu, C, Feng, L, Hardy, RF. and Hochberg, EJ, 2015. Spectral and spatial requirements of remote measurements of pelagic Sargassum macroalgae. *Remote Sensing of Environment*, [e–journal] 167. 229–246. http://dx.doi. org/10.1016/j.rse.2015.05.022.

Humpert, M. and Raspotnik, A, 2012. The Future of Arctic Shipping Along the Transpolar Sea Route. *Arctic Yearbook*, [e–journal] 2012. 281–307. Available at: <https://arcticyearbook.com/images/yearbook/2012/ Scholarly_Papers/14. Humpert_and_Raspotnik. pdf> [Accessed March 3, 2021].

Hussin, J, 2019. High–fat diet made Inuits healthier but shorter thanks to gene mutations, study finds. *The Controversation*, [blog] 26 March Available at: <https://theconversation.com/high– fat-diet-made-inuits-healthier–but–shorter–thanks–to–gene–mutations–study–finds–47529> [Accessed March 10, 2021].

IMGBIN.com, 2021. *Arctic Council Wine Logo Malbec PNG.* [online]. Available at: <https://imgbin.com/png/5HDK4Ee0/arctic–council–wine-logo–malbec–png> [Accessed March 10, 2021].

International Maritime Organization, 1989. *Coast Guard: Enforcement Under MARPOL V Convention on Pollution Expanded, Although Problems Remain.* [online]. Available at: <https://www.govinfo.gov/content/pkg/ GAOREPORTS-RCED-95-143/html/GAOREPORTS- RCED-95-143. htm> [Accessed March 3, 2021].

International Maritime Organization, 2019. *Convention on the Prevention of Marine Pollution by Dumping of Wastes and Other Matter 1972.* [online]. Available at: <https://www.imo.org/en/OurWork/Environment/Pages/ London–Convention–Protocol.aspx> [Accessed March 3, 2021].

International Union for Conservation of Nature, 2020. *'Plastic Waste–Free Islands' sounds like a dream, but a project works to bring it to reality.* [online]. Available at: <https://www.iucn.org/news/marine–and–polar/202002/ plastic–waste–free–islands–sounds–a–dream–a–project–works–bring-it–reality> [Accessed March 3, 2021].

International Whaling Commision, 2019. *A Joint IWC–IUCN–ACCOBAMS Workshop to Evaluate How the Data and Process Used to Identify Important Marine Mammal Areas (IMMAs) Can Assist the IWC to Identify Areas of High Risk for Ship Strike.* [online]. Available at: <https://www. marinemammalhabitat.org/download/report–of–the–joint–iwc–iucn-accobams– workshop–to–evaluate–how–the–data–and–process–used–to–identify–immas–can–assiste–the– iwc–to–identify–areas–of–high-risk–for–ship–strike/> [Accessed March 3, 2021].

Israel Space Agency, 2014. *The Shalom Mission*. [online]. Available at: <https://www.space.gov.il/en/research-and-development/1144> [Accessed March 3, 2021].

Jambeck, JR, Geyer, R, Wilcox, C, Siegler, TR, Perryman, M, Andrady, A, Narayan, R and Law, KL, 2015. Marine pollution. Plastic waste inputs from land into the ocean. *Science*, [e-journal] 347(6223). 768–771. http://dx.doi.org/10.1126/science.1260352.

Japan Agency for Marine-Earth Science and Technology, 2021. *Technology development for ocean resources exploration: Next-generation technology for ocean resources exploration*. [online]. Available at: <https://www.jamstec.go.jp/sip/en/enforcement-2/index.html> [Accessed March 10, 2021].

Jedlovec, G, 2009. Automated Detection of Clouds in Satellite Imagery. In: G. Jedlovec ed. *Advances in Geoscience and Remote Sensing*. InTech. http://dx.doi.org/10.5772/8326.

Kako, S, Morita, S and Taneda, T, 2020. Estimation of plastic marine debris volumes on beaches using unmanned aerial vehicles and image processing based on deep learning. *Marine Pollution Bulletin*, [e-journal] 155(28), 111127. http://dx.doi.org/10.1016/j. marpolbul.2020.111127.

Kaminski, C, Crees, T, Ferguson, J, Forrest, A, Williams, J, Hopkin, D and Heard, G, 2010. 12 days under ice – an historic AUV deployment in the Canadian High Arctic. In: *2010 IEEE/OES Autonomous Underwater Vehicles*. Monterey, CA, USA: IEEE. 1–11. http://dx.doi.org/10.1109/AUV.2010.5779651.

Kane, IA, Clare, MA, Miramontes, E, Wogelius, R, Rothwell, JJ, Garreau, and Pohl, F, 2020. Seafloor microplastic hotspots controlled by deep-sea circulation. *Science*, [e-journal] 368(6495). 1140–1145. http://dx.doi.org/10.1126/science.aba5899.

Karasik, R, Vegh, ZD, Bering, J, Caldas, J, Pickle, A, Rittschof, D and Virdin, J, 2020. *20 Years of Government Responses to the Global Plastic Pollution Problem The Plastics Policy Inventory*. [online]. Durham, NC. Available at: <https://nicholasinstitute.duke.edu/sites/default/ files/publications/20-Years-of-Government-Responses-to-the-Global-Plastic-Pollution-Problem-New_1.pdf> [Accessed March 3, 2021].

Katz, C, 2019. *Why Does the Arctic Have More Plastic than Most Places on Earth?* [online]. Available at: <https://www.nationalgeographic.com/science/2019/10/remote-arctic-contains-more-plastic-than-most-places-on-earth/> [Accessed March 3, 2021].

Kaye, S. and Warner, R, 2018. *Australian National Centre for Ocean Resources & Security (ANCORS)*. [online]. Available at: <https://www. uow.edu.au/ancors/#:%5Ctextasciitilde%20: text=ANCORS%20is%20 the%20only%20multidisciplinary,security%2C%20and%20marine% 20resources%20management.> [Accessed March 3, 2021].

Khoshhesab, Z. M, 2012. Reflectance IR Spectroscopy. In: T. Theophanides editor, *Infrared Spectroscopy*. London, UK: InTech. 233–244. http://dx.doi. org/10.5772/37180.

KIMO International, 2019a. *Fishing for Litter Fleet Cleans up after MSC Zoe but Who Pays the Costs?* [online]. Available at: <https://www. kimointernational.org/news/msc–zoe–who–pays–the–cleanup–costs/> [Accessed March 3, 2021].

KIMO International, 2019b. *Recycling Washed up Plastic Is More Challenging than You May Think*. [online]. Available at: <https://www. kimointernational.org/feature/recycling–washed–up–plastic–is–more– challenging–than–you–may–think/> [Accessed March 3, 2021].

Kirchner, S, 2017. The Future of the Central Arctic Ocean: Protection Through International Law. *The Journal of Territorial and Maritime Studies*, [e–journal] 4(2). 135–139.

Kirk, E, 2016. *Whose Job Is It to Protect the Arctic?* [online]. Available at: <http://theconversation. com/whose–job–is–it–to–protect–the– arctic–64778> [Accessed March 3, 2021].

Klein, S, Dimzon, IK, Eubeler, J and Knepper, T P, 2018. Analysis, Occurrence, and Degradation of Microplastics in the Aqueous Environment. [e– journal] 58. 51–67. http://dx. doi.org/10.1007/978–3–319–61615–5{\ textunderscore}3.

Klenke, M and Schenke, HW, 2002. A New Bathymetric Model for the Central Fram Strait. *Marine Geophysical Researches*, [e–journal] 23(4). 367–378. http://dx.doi.org/10.1023/A: 1025764206736.

Kühn, S, Bravo Rebolledo, EL. and van Franeker, JA, 2015. Deleterious Effects of Litter on Marine Life. *Marine Anthropogenic Litter*. 75–116. http://dx.doi.org/10.1007/978–3–319– 16510–3{\textunderscore}4.

Kuperman, J, 2014. The Northern Sea Route: Could It Be The New Suez Canal? Berkeley Political Review. *Berkeley Political Review*, [online]. Available at: <https://bpr.berkeley.edu/2014/12/22/the–northern–sea– route–the–new–suez–canal/> [Accessed March 3, 2021].

Kylin, H, 2020. Marine debris on two Arctic beaches in the Russian Far East. *Polar Research*, [e–journal] 39(0). http://dx.doi.org/10.33265/polar. v39.3381.

La Kanhai, DK, Gardfeldt, K, Krumpen, T, Thompson, RC and O'Connor, I, 2020. Microplastics in sea ice and seawater beneath ice floes from the Arctic Ocean. *Scientific Reports,* [e-journal] 10(1), 5004. http://dx.doi. org/10.1038/s41598-020-61948-6.

Larson, RW, 2014. *Disruptive innovation and naval power : strategic and financial implications of unmanned underwater vehicles (UUVs) and long-term underwater power sources.* [pdf]. Available at: <https://dspace.mit. edu/handle/1721.1/87959> [Accessed March 3, 2021].

Law, KL, 2017. Plastics in the marine environment. *Annual review of marine science,* [e-journal] 9. 205–229. http://dx.doi.org/10.1146/annurev-marine-010816-060409.

Law, KL, Morét-Ferguson, S, Maximenko, NA, Proskurowski, G, Peacock, EE, Hafner, J and Reddy, CM, 2010. Plastic accumulation in the North Atlantic subtropical gyre. *Science,* [e-journal] 329(5996). 1185–1188. http://dx.doi.org/10.1126/science.1192321.

Law, KL, Morét-Ferguson, SE, Goodwin, DS, Zettler, ER, Deforce, E, Kukulka, T and Proskurowski, G, 2014. Distribution of surface plastic debris in the eastern Pacific Ocean from an 11–year data set. *Environmental Science & Technology,* [e-journal] 48(9). 4732–4738. http://dx.doi.org/10.1021/es4053076.

Lebreton, L, Greer, SD. and Borrero, JC, 2012. Numerical modelling of floating debris in the world's oceans. *Marine Pollution Bulletin,* [e-journal] 64(3). 653–661. http://dx.doi.org/ 10.1016/j.marpolbul.2011.10.027.

Lebreton, L, Slat, B, Ferrari, F, Sainte-Rose, B, Aitken, J, Marthouse, R, Hajbane, S, Cunsolo, S, Schwarz, A, Levivier, A, Noble, K, Debeljak, P, Maral, H, Schöneich-Argent, R, Brambini, R, Reisser, J, Lebreton, L, Slat, B, Ferrari, F, Sainte-Rose, B, Hajbane, S, Cunsolo, S, Schwarz, A, Noble, K, Debeljak, P, Schoeneich-Argent, R, Brambini, R. and Reisser, J, 2018. Evidence that the Great Pacific Garbage Patch is rapidly accumulating plastic. *Scientific Reports,* [e-journal] 2018 // 8(1), 4666. http://dx.doi. org/10.1038/s41598– 018-22939-w.

Lebreton, L, van der Zwet, J, Damsteeg, J-W, Slat, B, Andrady, A and Reisser, J, 2017. River plastic emissions to the world's oceans. *Nature communications,* [e-journal] 8. 1–10. http://dx.doi.org/10.1038/ncomms15611.

Lee, S, Im, J, Kim, J, Kim, M, Shin, M, Kim, H.-c. and Quackenbush, L, 2016. Arctic Sea Ice Thickness Estimation from CryoSat-2 Satellite Data Using Machine Learning–Based Lead Detection. *Remote Sensing,* 8 [e-journal] (9), 698. http://dx.doi.org/10.3390/rs8090698.

Leeson, C, 2016. *A Plastic Ocean.* movie.

Lehning, H, 2018. *Qu'est-ce que la force de Coriolis ? D'où vient-elle et pourquoi fait-elle dévier les obus ?* [online]. Available at: <https://www. futura-sciences.com/sciences/questions-        reponses/mathematiques-fonctionne-force-coriolis-9671/> [Accessed March 3, 2021].

Lei, R. and Wei, Z, 2020. Exploring the Arctic Ocean under Arctic amplification. *Acta Oceanologica Sinica,* [e-journal] 39(9). 1-4. http:// dx.doi.org/10.1007/s13131-020-1642-9.

Leonardo Company, 2019. *PRISMA.* [online]. Available at: <https://www. leonardocompany. com/en/products/prisma> [Accessed March 3, 2021].

Lewis, D, 2016. *Are We Living in the Plastic Age?* [online]. Available at: <https://www.      smithsonianmag.com/smart-news/are-we-living-plastic-age-180957817/> [Accessed March 3, 2021].

Lewis, N, Riddle, MJ and Smith, SD, 2005. Assisted passage or passive drift: a comparison of alternative transport mechanisms for non-indigenous coastal species into the Southern Ocean. *Antarctic Science,* [e-journal] 17(2). 183-191. http://dx.doi.org/10.1017/S0954102005002580.

Li, WC, Tse, HF. and Fok, L, 2016. Plastic waste in the marine environment: A review of sources, occurrence and effects. *The Science of the total environment,* [e-journal] 566-567. 333-349. http://dx.doi.org/10.1016/j. scitotenv.2016.05.084.

Liedermann, M, Gmeiner, P, Pessenlehner, S, Haimann, M, Hohenblum, and Habersack, H, 2018. A Methodology for Measuring Microplastic Transport in Large or Medium Rivers. *Water,* [e-journal] 10(4), 414. http://dx.doi.org/10.3390/w10040414.

Lozoya, JP, Teixeira de Mello, F, Carrizo, D, Weinstein, F, Olivera, Y, Cedrés, F, Pereira, M. and Fossati, M, 2016. Plastics and microplastics on recreational beaches in Punta del Este (Uruguay): Unseen critical residents? *Environmental pollution (Barking, Essex : 1987),* [e-journal] 218. 931-941. http://dx.doi.org/10.1016/j.Envpol.2016.08.041.

Lu, L, Hang, D. and Di, L, 2015. Threshold model for detecting transparent plastic-mulched landcover using moderate-resolution imaging spectroradiometer time series data: a case study in southern Xinjiang, China. *Journal of Applied Remote Sensing,* [e-journal] 9(1), 097094. http:// dx.doi.org/10.1117/1.JRS.9.097094.

Lüber, S and McLellan, F, 2017. *Marine Debris and the Sustainable Development Goals.* [online]. Available at: <https://www.oceancare.org/ wp-content/uploads/2017/05/Marine_Debris_ neutral_2018_web.pdf> [Accessed March 3, 2021].

Ma, B, Steele, M and Lee, CM, 2017. Ekman circulation in the Arctic Ocean: Beyond the Beaufort Gyre. *Journal of Geophysical Research: Oceans,* [e-journal] 122(4). 3358–3374. http://dx.doi.org/10.1002/2016JC012624.

Maes, C, Grima, N, Blanke, B, Martinez, E, Paviet–Salomon, T and Huck, T, 2018. A Surface 'Superconvergence' Pathway Connecting the South Indian Ocean to the Subtropical South Pacific Gyre. *Geophysical Research Letters,* [e-journal] 45(4). 1915–1922. http://dx.doi.org/ 10.1002/2017GL076366.

Maes, T, van der Meulen, MD, Devriese, LI, Leslie, HA, Huvet, A, Frère, L, Robbens, J and Vethaak, AD, 2017. Microplastics Baseline Surveys at the Water Surface and in Sediments of the North–East Atlantic. *Frontiers in Marine Science,* [e–journal] 4. http://dx.doi.org/10.3389/ fmars.2017.00135.

Mahajan, S. and Fataniya, B, 2020. Cloud detection methodologies: variants and development—a review. *Complex & Intelligent Systems,* [e-journal] 6(2). 251–261. http://dx.doi.org/10. 1007/s40747–019–00128–0.

Mannaart, M, Bentley, A, McCord, G and Midavaine, J, 2019. *Marine Litter at UNESCO World Heritage Marine Sites.* [online]. Available at: <https:// www.kimonederlandbelgie.org/wp– content/uploads/FINAL_REPORT_ MARINE_LITTER_WHO.pdf> [Accessed March 3, 2021].

Maritime Research Institute Netherlands, 2020. *MARIN – Follow–up Research Container Loss Wadden.* [online]. Available at: <https://www. youtube.com/watch?v=dgxe7S1AXEY> [Accessed March 3, 2021].

Marlow, J, 2009. Report From Antarctica: Heaps of Trash or Historical Treasures? *Wired,* [blog] 08 April. Available at: <https://www.wired. com/2009/04/antarctictrash/> [Accessed March 11, 2021].

Martínez–Vicente, V, Clark, JR, Corradi, P, Aliani, S, Arias, M, Bochow, M, Bonnery, G, Cole, M, Cózar, A, Donnelly, R, Echevarría, F, Galgani, F, Garaba, SP, Goddijn–Murphy, L, Lebreton, L, Leslie, HA, Lindeque, K, Maximenko, N, Martin–Lauzer, F–R, Moller, D, Murphy, P, Palombi, L, Raimondi, V, Reisser, J, Romero, L, Simis, SG, Sterckx, S, Thompson, RC, Topouzelis, K N, van Sebille, E, Veiga, JM and Vethaak, AD,

Measuring Marine Plastic Debris from Space: Initial Assessment of Observation Requirements. *Remote Sensing,* [e–journal] 11(20), 2443. http://dx.doi.org/10.3390/rs11202443. Available at: <https: //www.mdpi. com/2072–4292/11/20/2443/htm> [Accessed March 3, 2021].

Masoumi, H, Safavi, SM. and Khani, Z, 2012. Identification and Classification of Plastic Resins Using near Infrared Reflectance Spectroscopy. *International Journal of Mechanical and Industrial Engineering,* (6). 213–220. Available at: <https://www.researchgate.net/ publication/285330830_ Identification_and_classification_of_plastic_resins_using_near_ infrared_reflectance_spectroscopy> [Accessed March 3, 2021].

Maximenko, N, Corradi, P, Law, KL, van Sebille, E, Garaba, SP, Lampitt, RS, Galgani, F, Martinez-Vicente, V, Goddijn-Murphy, L, Veiga, JM, Thompson, RC, Maes, C, Moller, D, Löscher, C. R, Addamo, AM, Lamson, MR, Centurioni, LR, Posth, NR, Lumpkin, R, Vinci, M, Martins, AM, Pieper, CD, Isobe, A, Hanke, G, Edwards, M, Chubarenko, I. P, Rodriguez, E, Aliani, S, Arias, M, Asner, GP, Brosich, A, Carlton, JT, Chao, Y, Cook, A.-M, Cundy, AB, Galloway, TS, Giorgetti, A, Goni, GJ, Guichoux, Y, Haram, LE, Hardesty, BD, Holdsworth, N, Lebreton, L, Leslie, HA, Macadam-Somer, I, Mace, T, Manuel, M, Marsh, R, Martinez, E, Mayor, DJ, Le Moigne, M, Molina Jack, ME, Mowlem, MC, Obbard, RW, Pabortsava, K, Robberson, B, Rotaru, A-E, Ruiz, GM, Spedicato, M. T, Thiel, M, Turra, A and Wilcox, C, 2019. Toward the Integrated Marine Debris Observing System. *Frontiers in Marine Science,* [e-journal] 6, 447. http://dx.doi.org/10.3389/fmars.2019.00447.

Mazzucato, M, 2021. *From Moonshots to Earthshots | by Mariana Mazzucato.* [online]. Available at: <https://www.project-syndicate.org/commentary/ moonshots-earthshots-state-investment-in-the-public-interest-by- mariana-mazzucato-2021-02> [Accessed March 3, 2021].

Merrell, TR, 1980. Accumulation of plastic litter on beaches of Amchitka Island, Alaska. *Marine Environmental Research,* [e-journal] 3(3). 171–184. http://dx.doi.org/10.1016/0141- 1136(80)90025-2.

Michel, APM, Morrison, AE, Preston, V. L, Marx, CT, Colson, BC. and White, H. K, 2020. Rapid Identification of Marine Plastic Debris via Spectroscopic Techniques and Machine Learning Classifiers. *Environmental Science & Technology,* [e-journal] 54(17). 10630–10637. http://dx.doi.org/10.1021/ acs.est.0c02099.

Migwi, FK, Ogunah, JA. and Kiratu, JM, 2020. Occurrence and Spatial Distribution of Microplastics in the Surface Waters of Lake Naivasha, Kenya. *Environmental Toxicology and Chemistry,* [e-journal] 39(4). 765–774. http://dx.doi.org/10.1002/etc.4677.

Milosevic, M, 2004. Internal Reflection and ATR Spectroscopy. *Applied Spectroscopy Reviews,* [e-journal] 39(3). 365–384. http://dx.doi. org/10.1081/ASR-200030195.

Misachi, J, 2017. *What Is Albedo?* [online]. Available at: <https://www. worldatlas.com/what-is- albedo.html> [Accessed March 3, 2021].

Monitoring, A. and Program, A, 2018. *Biological Effects of the contaminats on Arctic wildlife & Fish. Summary for Policy-Makers.* [online]. Available at: <https://www.amap.no/documents/ download/3297/inline> [Accessed March 10, 2021].

Moore, C, Moore, S, Leecaster, M. and Weisberg, S, 2001. A Comparison of Plastic and Plankton in the North Pacific Central Gyre. *Marine Pollution Bulletin,* [e-journal] 42(12). 1297–1300. http://dx.doi.org/10.1016/S0025-326X(01)00114-X.

Morishige, C, 2012. Garbage Patches: The Cost of a Cleanup (Part 2) | OR&R's Marine Debris Program. *NOAA Marine Debris Program Blog,* [blog] 13 July. Available at: <https://blog.marinedebris.noaa.gov/garbage–patches–cost–cleanup–part–2> [Accessed March 3, 2021].

Morishige, C, Donohue, MJ, Flint, E, Swenson, C and Woolaway, C, 2007. Factors affecting marine debris deposition at French Frigate Shoals, Northwestern Hawaiian Islands Marine National Monument, 1990–2006. *Marine Pollution Bulletin,* [e-journal] 54(8). 1162–1169. http://dx.doi.org/10.1016/j.marpolbul.2007.04.014.

Moy, K, Neilson, B, Chung, A, Meadows, A, Castrence, M, Ambagis, S. and Davidson, K,

Mapping coastal marine debris using aerial imagery and spatial analysis. *Marine Pollution Bulletin,* [e-journal] 132. 52–59. http://dx.doi.org/10.1016/j.marpolbul.2017.11. 045.

Nansen, F, 1897. *Farthest North.* New York: Archibald Constable and Company.

National Aeronautics and Space Administration, 2004. *Novaya Zemlya, Northern Russia.* [online]. Available at: <https://visibleearth.nasa.gov/images/107641/novaya–zemlya–northern– russia> [Accessed March 10, 2021].

National Aeronautics and Space Administration (NASA) Global Climate, 2021. *Arctic Sea Ice Minimum | Vital Signs – Climate Change: Vital Signs of the Planet.* [online]. Available at: <https://climate.nasa.gov/vital–signs/arctic–sea–ice/> [Accessed March 11, 2021].

National Oceanic and Atmospheric Administration, 2011. *Ocean Currents.* [online]. Available at: <https://www.noaa.gov/education/resource-collections/ocean–coasts/ocean–currents> [Accessed March 3, 2021].

National Oceanic and Atmospheric Administration, 2013. *Garbage Patches OR&R's Marine Debris Program.* [online]. Available at: <https://marinedebris.noaa.gov/info/patch.html> [Accessed March 3, 2021].

National Oceanic and Atmospheric Administration, 2020a. FY 2021 Grant Opportunity for North America Marine Debris Prevention and Removal Projects. *NOAA Marine Debris Program Blog,* [blog] 10 Nov. Available at: <https://blog.marinedebris.noaa.gov/now-open-fy-2021–grant-opportunity–north–america–marine–debris–prevention–and–removal-projects> [Accessed March 3, 2021].

National Oceanic and Atmospheric Administration, 2020b. *Thermohaline Circulation – Currents: NOAA's National Ocean Service Education.* [online]. Available at: <https://oceanservice.noaa. gov/education/ tutorial_currents/05conveyor1.html> [Accessed March 3, 2021].

National Oceanic and Atmospheric Administration, 2020c. *What Is a Gyre?* [online]. Available at: <https://oceanservice.noaa.gov/facts/gyre.html> [Accessed March 3, 2021].

National Oceanic and Atmospheric Administration, 2021. *Why Do Scientists Measure Sea Surface Temperature?* [online]. Available at: <https://oceanservice.noaa.gov/facts/sea–surface– temperature.html> [Accessed March 3, 2021].

National Oceanic and Atmospheric Administration Marine Debris Program, 2020. *NOAA Marine Debris Program FY 2021–2025 Strategic Plan.* Silver Spring, MD: National Oceanic and Atmospheric Administration Marine Debris Program.

National Snow and Ice Data Center, 2011. *Arctic sea ice at record low for Julyy | Arctic Sea Ice News and Analysis.* [online]. Available at: <https://nsidc.org/arcticseaicenews/2011/08/arctic– sea–ice–at–record–low–for–Julyy/> [Accessed March 11, 2021].

National Snow and Ice Data Center, 2020a. *All About Sea Ice.* [online]. Available at: <https://nsidc.org/cryosphere/seaice/index.html> [Accessed March 3, 2021].

National Snow and Ice Data Center, 2020b. *Arctic vs. Antarctic.* [online]. Available at: <https://nsidc.org/cryosphere/seaice/characteristics/ difference.html> [Accessed March 3, 2021].

National Snow and Ice Data Center, 2020c. *Thermodynamics: Albedo.* Available at: <https://nsidc.org/cryosphere/seaice/processes/albedo. html>.

Navingo, 2018. *ISE Explorer AUV for COMRA.* [online]. Available at: <https://www.offshore–energy.biz/ise–explorer–auv–for–comra/> [Accessed March 3, 2021].

Naware, R, 2018. *The Mediterranean at Risk of Becoming 'a Sea of Plastic', WWF Warns | WWF.* [online]. Available at: <https://wwf.panda.org/ wwf_news/press_releases/?329099/The–Mediterranean–at–risk–of– becoming–a–sea–of–plastic–WWF–warns> [Accessed March 3, 2021].

Newman, S, Watkins, E, Farmer, A, Brink, ten and Schweitzer, J.–P, 2015. The Economics of Marine Litter. In: M. Bergmann, L. Gutow and M. Klages eds. *Marine Anthropogenic Litter.* Cham: Springer International Publishing, Imprint, and Springer. 367–394. http://dx.doi.org/10.1007/978–3–319– 16510–3{\textunderscore}14.

Niaounakis, M, 2017. *Management of Marine Plastic Debris*. Plastics Design Library. Saint Louis: Elsevier Science.

Nielsen, TD, Hasselbalch, J, Holmberg, K. and Stripple, J, 2020. Politics and the plastic crisis: A review throughout the plastic life cycle. *WIREs Energy and Environment*, [e–journal] 9(1). http://dx.doi.org/10.1002/wene.360.

Nordregio, 2019. *Sea Routes and Ports in the Arctic*. [online]. Available at: <https://nordregio.    org/maps/sea–routes–and–ports–in–the–arctic/> [Accessed March 3, 2021].

Obbard, R. W, Sadri, S, Wong, YQ, Khitun, AA, Baker, I. and Thompson, RC, 2014. Global warming releases microplastic legacy frozen in Arctic Sea ice. *Earth's Future*, [e–journal] 2(6). 315–320. http://dx.doi.org/10.1002/2014EF000240.

Ocean Cleanup Project, 2021. *Oceans*. [online]. Available at: <https://theoceancleanup.com/ oceans/> [Accessed March 3, 2021].

Omerdic, E and Toal, D, 2007. Modelling of waves and ocean currents for real–time simulation of ocean dynamics. In: *Oceans 2007 – Europe*. Piscataway, NJ: IEEE Service Center. 1–6. http://dx.doi.org/10.1109/OCEANSE.2007.4302323.

ONDA, 2020a. *Copernicus Sentinel–1 Mission – ONDA DIAS*. [online]. [Accessed March 3, 2021].

ONDA, 2020b. *Copernicus Sentinel–3 Mission – ONDA DIAS*. [online]. Available    at:    <https://www.onda–dias.eu/cms/de/data/catalogue/sentinel–3/> [Accessed March 3, 2021].

Organization, IM, 1988. *International Convention for the Prevention of Pollution from Ships (MARPOL)*. [online]. Available at: <https://www.imo.org/en/About/Conventions/Pages/   International–Convention–for–the–Prevention–of–Pollution–from–Ships–(MARPOL).aspx> [Accessed March 3, 2021].

Ota, Y, 2017. For indigenous communities, fish mean much more than food. *bifrostonline.org*, [blog] 29 January. Available at: <https://bifrostonline.org/for–indigenous–communities–fish–mean–much–more–than–food/> [Accessed March 3, 2021].

Ouadih, AE, Ouladsaid, A. and Hassani, MM, 2011. *Antenna in Relation to the Penetration Depth of the Wave in Water Used for Communication with the Submarine*. [online]. Available at: <https://www.researchgate.net/publication/266087975_Antenna_in_relation_to_the_ penetration_depth_of_the_wave_in_water_used_for_communication_with_the_submarine> [Accessed March 3, 2021].

Pahl, S. and Wyles, KJ, 2017. The human dimension: how social and behavioural research methods can help address microplastics in the environment. *Analytical Methods*, [e-journal] 9(9). 1404–1411. http://dx.doi.org/10.1039/C6AY02647H.

Peeken, I, Primpke, S, Beyer, B, Gütermann, J, Katlein, C, Krumpen, T, Bergmann, M, Hehemann, L and Gerdts, G, 2018. Arctic sea ice is an important temporal sink and means of transport for microplastic. *Nature communications*, 9 [e-journal](1), 1505. http: //dx.doi.org/10.1038/s41467–018–03825–5.

Peeters, 2021. *Interview with the team about the project direction.* [interview]. Available at: <https://docs.google.com/ document/d/1PmInUZeq35wnG 9tfyMfZDUnK1fDiVTCO/edit> [Accessed March 3, 2021].

Pfirman, SL, Kogeler, J. and Anselme, B, 1995. Coastal environments of the western Kara and eastern Barents Seas. *Deep Sea Research Part II: Topical Studies in Oceanography,* [e-journal] 42(6). 1391–1412. http://dx.doi.org/10.1016/0967–0645(95)00047–X.

Polasek, L, Bering, J, Kim, H, Neitlich, P, Pister, B, Terwilliger, M, Nicolato, K, Turner, C and Jones, T, 2017. Marine debris in five national parks in Alaska. *Marine Pollution Bulletin,* [e-journal] 117(1–2). 371–379. http://dx.doi.org/10.1016/j.marpolbul.2017.01.085.

PortandTerminal.com, 2020. *Are containers lost overboard pollution? Yang Ming ship arrested in landmark case.* [online]. Available at: <https://www.portandterminal.com/why–is–yang ming–refusing–to–clean–up–its–mess–in–australia/> [Accessed March 3, 2021].

Prata, JC, 2018. Airborne microplastics: Consequences to human health? *Environmental pollution (Barking, Essex : 1987),* [e-journal] 234. 115–126. http://dx.doi.org/10.1016/j.envpol.2017.11.043.

Prisma, 2021. *PRISMA: Small Innovative Earth Observation Mission.* [online]. Available at: <http://prisma-i.it/index.php/en/> [Accessed March 3, 2021].

Protection of the Arctic Marine Environment, 2020. *The increase in Arctic shipping 2013-2019.* Available at: <https://www.pame.is/projects/arctic–marine–shipping/arctic–shipping–status–reports/723–arctic–shipping–report–1–the–increase–in–arctic–shipping–2013–2019–pdf–version/file>.

Protection of the Arctic Marine Environment, 2021. *Part IIA – Pollution Prevention Measures – Chapter 5.* [online]. Available at: <https://www.pame.is/part–iia–pollution–prevention– measures–chapter–5> [Accessed March 3, 2021].

Puiu, T, 2017. *The Types of Clouds: Everything You Need to Know.* [online]. Available at: <https://www.zmescience.com/science/types-of-clouds/> [Accessed March 3, 2021].

Quinn, E, 2019. Iceland to host international symposium on plastics in Arctic and sub–Arctic – Eye on the Arctic. *Eye on the Arctic,* [blog] 25 October. Available at: <https://www.rcinet.ca/eye- on-the–arctic/2019/10/21/ iceland–to–host–international–symposium–on–plastics–in–arctic- and–sub–arctic/> [Accessed March 3, 2021].

Rahmstorf, S, 2006. Thermohaline Circulation – Fact Sheet by Stefan Rahmstorf. In: SA. Elias edited by *Encyclopedia of Quaternary Sciences.* Amsterdam: Elsvier. Available at: <http://www.pik- potsdam.de/~stefan/ Publications/Book_chapters/rahmstorf_eqs_2006.pdf>.

Riskas, K, 2019. The Indian Ocean's Great Disappearing Garbage Patch. *Hakai Magazine,* [online]. Available at: <https://www.hakaimagazine. com/news/the–indian–oceans–great–disappearing–garbage–patch/> [Accessed March 3, 2021].

Rochman, CM, Kurobe, TFlores, I and Teh, SJ, 2014. Early warning signs of endocrine disruption in adult fish from the ingestion of polyethylene with and without sorbed chemical pollutants from the marine environment. *The Science of the total environment,* [e–journal] 493. 656–661. http:// dx.doi.org/10.1016/j.scitotenv.2014.06.051.

Ryan, C, Thomas, G. and Stagonas, D, 2020. Arctic Shipping Trends 2050. *s.l.* http://dx.doi. org/10.13140/RG.2.2.34680.67840.

Ryan, G, Dilley, BJ, Ronconi, RA and Connan, M, 2019. Rapid increase in Asian bottles in the South Atlantic Ocean indicates major debris inputs from ships. *Proceedings of the National Academy of Sciences,* [e–journal] 116(42). 20892–20897. http://dx.doi.org/10. 1073/pnas.1909816116.

Saint–Raymond, L, 2010. *Gyres océaniques et ondes de Rossby piégées.* [pdf]. Paris. Available at: <https:// www.college–de–france.fr/media/pierre- louis–lions/UPL37046_Laure_Saint_Raymond.pdf>.

Santos, IR, Friedrich, AC, Wallner–Kersanach, M. and Fillmann, G, 2005. Influence of socioeconomic characteristics of beach users on litter generation. *Ocean & Coastal Management,* [e–journal] 48(9–10). 742– 752. http://dx.doi.org/10.1016/j.Ocecoaman.2005.08.006.

Schmaltz, E, Melvin, EC, Diana, Z, Gunady, EF, Rittschof, D, Somarelli, JA, Virdin, J and Dunphy–Daly, MM, 2020. Plastic pollution solutions: emerging technologies to prevent and collect marine plastic pollution. *Environment International,* [e–journal] 144(106067). 1–17. http://dx.doi. org/10.1016/j.envint.2020.106067.

Schøyen, H. and Bråthen, S, 2011. The Northern Sea Route versus the Suez Canal: cases from bulk shipping. *Journal of Transport Geography,* [e- journal] 19(4). 977–983. http://dx.doi.org/10.1016/j.jtrangeo.2011.03.003.

Sebille, E. van, England, MH. and Froyland, G, 2012. Origin, dynamics and evolution of ocean garbage patches from observed surface drifters. *Environmental Research Letters,* [e–journal] 7(4), 044040. http://dx.doi.org/10.1088/1748–9326/7/4/044040.

Serranti, S, Gargiulo, A, Bonifazi, G, Toldy, A, Patachia, S. and Buican, R, 2010. The Utilization of Hyperspectral Imaging for Impurities Detection in Secondary Plastics. *The Open Waste Management Journal,* [e–journal] 3(1). 56–70. http://dx.doi.org/10.2174/ 1876400201003010056.

Shirah, G and Mithchell, H, 2015. *Garbage Patch Visualization Experiment.* [online]. Available at: <https://svs.gsfc.nasa.gov/4174> [Accessed March 3, 2021].

Silber, GK. and Adams, JD, 2019. Vessel Operations in the Arctic, 2015–2017. *Frontiers in Marine Science,* [e–journal] 6. i–194. http://dx.doi.org/10.3389/fmars.2019.00573.

Simmonds, I, 2015. Comparing and contrasting the behaviour of Arctic and Antarctic sea ice over the 35 year period 1979–2013. *Annals of Glaciology,* [e–journal] 56(69). 18–28. http://dx.doi.org/10.3189/2015AoG69A909.

Simon, M, 2019. *The Riddle, and Controversy, of All That Missing Plastic.* [online]. Wired. Available at: <https://www.wired.com/story/missing–plastic/> [Accessed March 3, 2021].

Singh, K, Bjerregaard, and Man Chan, H, 2014. Association between environmental contaminants and health outcomes in indigenous populations of the Circumpolar North. *International Journal of Circumpolar Health,* [e–journal] 73(1), 25808. http://dx.doi.org/10. 3402/ ijch.v73.25808.

Slat, B, Ardiyanti, A, Arens, E, Bolle, E, Hyke Brugman, Campbell, H, Pierre–Louis Christiane, Cooper, B, Dekker, M, Diamant, S, van Dijk, B, van Dijk, M, Drenkelford, S, Jabe Faber, Fransesco F. Ferrari, Fraunholcz, N, Geil, JC, Göbel, C, Grassini, D, Haan, J. de, Hammer, J, Hildreth, R. G, Tadzio Holtrop, Hornsby, E, van der Horst, S, Howard, N, Jansen, H, Jansen, W, Jonkman, JA, Katsepontes, NP, Kauzlaric, D, Klein, R, Koch, S, Landgraf, T, Leinenga, R, Limouzin, C, van Lith, EH, López, J, Wart Luscuere, Martini, R, Michels, M, Monteleone, B, Osborn, K, Ozmutlu, S, Pallares, E, Pavlov, L, Peek, H, Kauê Pelegrini, Prendergast, GS, Proietti, M, Randolph, J, Rau, A, Reisser, J, Roelofs, M, van Rossum, S, Rushton, R, van Ruyven, S, Sainte–Rose, B, Samir, S, van Schie, S, Schmid, M, Schmidt, K, Sivasubramanian, N, Stack, KL, Walker, P, Walters, A, Walworth, N, Wehkamp, N, Zuijderwijk, R and Sonneville, J de, 2014. *Feasibility Study – The Ocean Cleanup.* 2nd ed. The Netherlands: HOAX Graphic Design. Available at: <https://www.researchgate.net/publication/309202751_Feasibility_Study_–_The_Ocean_Cleanup> [Accessed March 3, 2021].

Stachowitsch, M, 2019. *The Beachcombers Guide to Marine Debris.* Cham: Springer International Publishing.

Starr, C, 2016. *Annual Arctic Sea Ice Minimum 1979-2015 with Area Graph.* [online]. Available at: <https://svs.gsfc.nasa.gov/4435> [Accessed March 3, 2021].

Statista, 2020. *Global plastic production 1950-2018 I Statista.* [online] Available at: <https://www.statista.com/statistics/282732/global-production-of-plastics-since-1950/>. [online]. [Accessed December. 16, 2020].

Szczepanski, M, 2019. *Study Puts Societal Cost on Ocean Plastic Pollution.* [online]. Available at: <https://www.waste360.com/plastics/study-puts-economic-social-cost-ocean-plastic- pollution> [Accessed March 3, 2021].

Tadesse, Y, Villanueva, A, Haines, C, Novitski, D, Baughman, R and Priya, S, 2012. Hydrogen- fuel-powered bell segments of biomimetic jellyfish. *Smart Materials and Structures,* [e-journal] 21(4), 045013. http://dx.doi.org/10.1088/0964-1726/21/4/045013.

Taipale, SJ, Peltomaa, E, Kukkonen, JVK, Kainz, MJ, Kautonen, and Tiirola, M, 2019. Tracing the fate of microplastic carbon in the aquatic food web by compound-specific isotope analysis. *Scientific Reports,* [e-journal] 9(1), 19894. http://dx.doi.org/10.1038/s41598-019- 55990-2.

Tan, S.-Y, 2020. *Interview with Team.* Available at: <https://docs.google.com/document/d/1iY984SY8h5E6EreiWSUuFoEIXoR9m5t_Cf7JUrIcSBU/edit?usp=sharing> [Accessed March 3, 2021].

Tekman, MB, Krumpen, T and Bergmann, M, 2017. Marine litter on deep Arctic seafloor continues to increase and spreads to the North at the HAUSGARTEN observatory. *Deep Sea Research Part I: Oceanographic Research Papers,* [e-journal] 120. 88-99. http://dx.doi.org/ 10.1016/j.dsr.2016.12.011.

The Great Bubble Barrier BV, 2020. *Bubble Barrier - Homepage.* [online]. Available at: <https://thegreatbubblebarrier.com/?lang=en> [Accessed March 3, 2021].

The Innoter Group of Companies, 2020a. *Sentinel-1A, 1B.* [online]. Available at: <https://innoter. com/en/satellites/sentinel-1a-1b/> [Accessed March 3, 2021].

The Innoter Group of Companies, 2020b. *TerraSAR-X/TanDEM-X.* [online]. Available at: <https://innoter.com/en/satellites/terrasar-x-tandem-x/> [Accessed March 3, 2021].

The Maritime Executive editor, 2021. *Video: Coastal Freighter Breaks Up at Anchor.* [online]. Available at: <https://maritime-executive.com/article/

video–coastal–freighter–breaks–up–at–anchor> [Accessed March 3, 2021].

Themistocleous, K, Papoutsa, C, Michaelides, S. and Hadjimitsis, D, 2020. Investigating Detection of Floating Plastic Litter from Space Using Sentinel–2 Imagery. *Remote Sensing*, [e–journal] 12(16), 2648. http:// dx.doi.org/10.3390/rs12162648.

Thevenon, F, Carroll, C and Sousa, J, 2014. *Plastic debris in the ocean: The characterization of marine plastics and their environmental impacts, situation analysis report.* Gland: IUCN International Union for Conservation of Nature and Natural Resources.

Tilling, RL, Ridout, A, Shepherd, A. and Wingham, DJ, 2015. Increased Arctic sea ice volume after anomalously low melting in 2013. *Nature Geoscience,* [e–journal] 8(8). 643646. http://dx.doi.org/10.1038/ngeo2489.

Topouzelis, K, Papakonstantinou, A and Garaba, SP, 2019. Detection of floating plastics from satellite and unmanned aerial systems (Plastic Litter Project 2018). *International Journal of Applied Earth Observation and Geoinformation,* [e–journal] 79. 175–183. http://dx.doi.org/ 10.1016/j. jag.2019.03.011.

Tramoy, R, Gasperi, J, Dris, R, Colasse, L, Fisson, C, Rocher, V. and Bruno, B, 2019. *Flux et dynamique de transfert des macroplastiques du continent à la mer: le cas de la Seine.* [online]. [pdf]. Available at: <https://po2019. sciencesconf.org/263597/document> [Accessed March 3, 2021].

Trishchenko, AP. and Garand, L, 2011. Spatial and Temporal Sampling of Polar Regions from Two–Satellite System on Molniya Orbit. *Journal of Atmospheric and Oceanic Technology,* [e–journal] 28(8). 977–992. http:// dx.doi.org/10.1175/JTECH–D–10–05013.1.

Tufts University, 2021. *Chapter 8: The Arctic & the LOSC Law of the Sea.* [online]. Available at: <https://sites.tufts.edu/lawofthesea/chapter–eight/> [Accessed March 3, 2021].

U.S. Census, 2021. *Population Clock.* [online]. Available at: <https://www. census.gov/popclock/> [Accessed March 11, 2021].

United Nations, 1982. *United Nations Convention on the Law of the Sea.* [online]. Available at: <https://www.un.org/depts/los/convention_ agreements/texts/unclos/unclos_e.pdf> [Accessed March 3, 2021].

United Nations, 2009. *State of the World's Indigenous Peoples.* Economic & social affairs. New York: United Nations. isbn: 978–92–1–130283–7.

United Nations Environment Program, 2017. *Marine Litter Socio Economic Study.* [online]. Available at: <https://wedocs.unep.org/ handle/20.500.11822/26014> [Accessed March 3, 2021].

University Corporation for Atmospheric Research, 2019. *Cloud Types | UCAR Center for Science Education.* [online]. Available at: <https://scied. ucar.edu/learning-zone/clouds/cloud- types> [Accessed March 3, 2021].

University of Lapland, 2021. *Arctic Indigenous Peoples.* [online]. Arctic Centre. Available at: <https://www.arcticcentre.org/EN/arcticregion/ Arctic-Indigenous-Peoples> [Accessed March 10, 2021].

van Vliet, A, 2013. *The Story of Capannori. A Zero Waste champion: Best Practices.* [pdf]. Available at: <https://zerowasteeurope.eu/wp-content/ uploads/2013/09/ZWE-Best-practice-Capannori.pdf>[Accessed December. 1, 2021].

Verny, J. and Grigentin, C, 2009. Container shipping on the Northern Sea Route. *International Journal of Production Economics,* [e-journal] 122(1). 107–117. http://dx.doi.org/10.1016/j. ijpe.2009.03.018.

Vessel Finder, 2019. *Careless loading of cargo containers sinks ship in Iran port (Video).* Ed. by Vessel Finder. [online]. Available at: <https://www. vesselfinder.com/news/14954-Careless-loading-of-cargo-containers- sinks-ship-in-Iran-port-Video> [Accessed March 3, 2021].

Vidal, J, 2010. *Modern Cargo Ships Slow to the Speed of the Sailing Clippers I Travel and Transport I The Guardian.* [online]. Available at: <https:// www.theguardian.com/environment/2010/July/25/slow-ships-cut- greenhouse-emissions> [Accessed March 3, 2021].

Vitale, A, 2010. *Il Confronto Internazionale Nell'Artico.* [online]. Available at: <https://www.parlamento.it/documenti/repository/affariinternazionali/ osservatorio/approfondimenti/Approfondimento_24_ISPI_Artico.pdf> [Accessed March 3, 2021].

Voytenko, M, 2021. *Ukrainian Cargo Ship Sank or Sinking at Bartin Anchorage, Black Sea.* [online]. Available at: <https://www.fleetmon.com/ maritime-news/2021/32305/ukrainian-cargo-ship-sank-or-sinking- bartin-anchor/> [Accessed March 3, 2021].

Wall, M, 2015. *World View Offers Cost-Sharing Balloon Flights to Stratosphere.* [online]. Available at: <https://www.space.com/30477-world-view- balloon-flights-cost-sharing.html> [Accessed March 3, 2021].

Watanabe, J-i, Shao, Y and Miura, N, 2019. Underwater and airborne monitoring of marine ecosystems and debris. *Journal of Applied Remote Sensing,* [e-journal] 13(04), 1. http: //dx.doi.org/10.1117/1.JRS.13.044509.

Wells, NC, 2015. *Oceanographic Topics | General Processes.*

Whitmire, SL and van Bloem, SJ, 2017. *Quantification of Microplastics on National Park Beaches.* [online]. Available at: <https://marinedebris. noaa.gov/sites/default/files/publications-files/Quantification_of_ Microplastics_on_National_Park_Beaches.pdf> [Accessed March 3, 2021].

Woods Hole Oceanographic Institution, 2021. *Arctic Ocean Circulation.* [online]. Available at: <https://www.whoi.edu/know–your–ocean/ocean-topics/polar–research/arctic–ocean– circulation/> [Accessed March 11, 2021].

World Shipping Council, 2017. *Containers Lost At Sea – 2017 Update.* [pdf]. Available at: <https://www.worldshipping.org/industry–issues/safety/Containers_Lost_at_Sea_–_2017_ Update_FINAL_July_10.pdf> [Accessed March 3, 2021].

World Shipping Council, 2020. *Containers Lost at Sea – 2020 Update.* [pdf]. Available at: <https://www.worldshipping.org/Containers_Lost_at_Sea_–_2020_Update_FINAL_.pdf> [Accessed March 3, 2021].

World View, 2021a. *World View.* [online]. Available at: <https://worldview.space/> [Accessed March 3, 2021].

World View, 2021b. *World View Product.* [online]. Available at: <https://worldview.space/product/> [Accessed March 3, 2021].

Young, OR, 2016. The Arctic Council at Twenty: How to Remain Effective in a Rapidly Changing Environment. *UC Irvine Law Review*, [online] 6(99). 99–120. Available at: <https://scholarship.law.uci.edu/ucilr/vol6/iss1/5> [Accessed March 3, 2021].

Zero 2 Infinity SL, 2016. *Zero 2 Infinity Media.* [online]. Available at: <https://www.zero2infinity.space/media/> [Accessed March 3, 2021].

Zero 2 Infinity SL, 2020. *From Andalusia to space: a zero–emissions odyssey.* [pdf]. Available at: <https://www.zero2infinity.space/in–the–news/andalusia–space–zero–emissions–odyssey– 2/> [Accessed March 8, 2021].

Zero 2 Infinity SL, 2021. *Zero 2 Infinity Company.* [online]. Available at: <https://www. zero2infinity.space/company/> [Accessed March 3, 2021].

Zhang, W, Xu, X. and Wu, Y, 2016. A new method of single celestial–body sun positioning based on theory of mechanisms. *Chinese Journal of Aeronautics*, [e–journal] 29(1). 248–256. http://dx.doi.org/10.1016/j.cja.2015.12.009.

Zhu, S, Chen, H, Wang, M, Guo, X, Lei, Y and Jin, G, 2019. Plastic solid waste identification system based on near infrared spectroscopy in combination with support vector machine. *Advanced Industrial and Engineering Polymer Research*, [e–journal] 2(2). 77–81. http://dx. doi.org/10.1016/j.aiepr.2019.04.001.

# A
# Appendix

---

Selection of satellites used for remote sensing to monitor the oceans. These satellites could be used to detect and track plastics in the oceans.

**Table 11:** *Satellites Monitoring Oceans. Image credits: 1. ESA, 2021, 2. ESA, 2021, 3. ESA, 2008, 4. ESA, 2020, 5. ONDA, 2020b, 6. World View, 2021b, 7. Airbus, 2021, 8. Prisma, 2021, 9. EnMAP Ground Segment Team, 2020, 10. Israel Space Agency, 2014*

| Satellite - Launch Year | Description | Image |
|---|---|---|
| 1. Sentinel 1A and 1B (2014 and 2016) | Manufactured by Thales Alenia Space and Airbus Defense Space (ESA, 2021; 2021; Innoter, 2020). Part of ESA Copernicus Program. Equipped with C-band SAR sensors (ONDA, 2020a). | |
| 2. Sentinel 2A and 2B (2015 and 2017) | Manufactured by EADS Ashium Satellites. Equipped with a multispectral instrument (ESA, 2019; 2021). | |
| 3. TanDEM-X and TerraSAR-X (2007 and 2010) | Manufactured by EADS Ashitun Satellites (Innoter, 2020). Operated by the German Space Agency, Deutsches Zentrum für Luft- und Raumfahrt. Both are equipped with X-band SAR sensors (ESA, 2008; 2020). | |

| 4. RADARSAT-2 (2017) | Manufactured by MDA Corporation. Launched in 2007 as a joint mission between the Canadian Space Agency and MDA Corporation. Equipped with a C-band SAR sensor. (Canadian Space Agency, 2021; ESA, 2020) | |
|---|---|---|
| 5. Sentinel 3A and 3B (2015 and 2017) | Manufactured by Thales Alenia Space. Operated by ESA and the European Organization for the Exploitation of Meteorological Satellites (EUMETSAT). Equipped with an Ocean and Land Color Instrument, Sea and Land Surface Temperature Radiometer (SLSTR), Synthetic Aperture Radar Altimeter, and Microwave Radiometer. (ESA, 2021; ONDA, 2020b). | |
| 6. WorldView-3 (2014) | Commercial satellite manufactured by Ball Aerospace and Technologies Corporation and operated by Digital Globe Inc.. Equipped with a WorldView-110 camera that collects images in 29 spectral bands. (ESA, 2021; World View, 2021a) | |
| 7. Neo Pleiades (2021) | Neo Pleiades is a vely high-resolution optical imagely satellite constellation that is manufactured and operated by Airbus Defence and Space (2021; 2021). | |

| 8. PRISMA (2019) | Funded and operated by the Italian Space Agency (ESA, 2019; 2020). It was manufactured by Leonardo Company, OHB Italia (Prisma, 2021), and is equipped with a hyperspectral imager (Leonardo Company, 2019). | |
|---|---|---|
| 9. EnMAP (2022) | EnMAP is an Earth observation satellite operated by the Deutsches Zentrum für Luft- und Raumfahrt (2021). It is manufactured by OHB System, Bremen and equipped with a hyperspectral instrument (ESA, 2017). | |
| 10. SHALOM (2021) | Earth observation satellite funded and operated by the Israel Space Agency and Agenzia Spaziale Italiana. It is manufactured by Elbit System, Leonardo Company and Thales Alenia Space. Equipped with a panchromatic camera with a 2.5m ground sample distance. (Israel Space Agency, 2014) | |

CPSIA information can be obtained
at www.ICGtesting.com
Printed in the USA
JSHW051150271022
32214JS00003B/104